FORENSIC SCIENCE

FORENSIC SCIENCE

How Scientific Evidence Has Transformed Criminal Trials

Sanjukta Mondal

ICON

This book is dedicated to the memory of my loving grandfather, who taught me that living is to keep learning, through people, through pages, through experiences good and bad, and especially through mistakes. He believed life does not end with a final breath, it ends the moment we cease to learn.

Published in the UK in 2026 by
Icon Books Ltd, Omnibus Business Centre,
39–41 North Road, London N7 9DP
email: info@iconbooks.com
www.iconbooks.com

ISBN: 978-183773-295-1
eBook: 978-183773-296-8

Typeset by SJmagic DESIGN SERVICES, India.

Printed and bound in the UK.

Appointed GPSR EU Representative:
Easy Access System Europe Oü, 16879218
Address: Mustamäe tee 50, 10621, Tallinn, Estonia
Contact Details: gpsr.requests@easproject.com, +358 40 500 3575

ACKNOWLEDGEMENTS

I don't want to be one of those recipe blogs that give you two pages of life updates before finally getting to the chocolate chip cookie recipe. So, here's a quick shoutout to the people who made sure this book was actually written.

To my biggest cheerleaders, Maa and Baba, you were right. Thank you for believing that your daughter would one day channel her fascination for forensic mysteries into something good instead of becoming a criminal. Well, here you go. I wrote a book! To my husband, Ayan, thank you for making sure I stuck to my schedule, didn't forget to eat, and didn't wander around the house like a sleep-deprived zombie deep in research. To my best friends, Taisha and Pragya, thank you for patiently listening to all the 'did you knows' at the end of long, writing-filled days and reminding me to trust my myself.

CONTENTS

INTRODUCTION

Imagine, for the sake of this argument, that you have successfully stolen a car. No witnesses, no fingerprints, and no one on your trail. You sit inside the vehicle, patting yourself on the back for committing the perfect crime, and that's when you feel the sting of a mosquito bite. You try to kill it, but it escapes. A few days later, the police come knocking at your door for stealing a car. How did they get here? The mosquito snitched on you. The blood sucked by the mosquito matches your DNA, proving that you were at the crime scene. Sounds like the plotline from one of many beloved forensic or crime scene investigation shows, right?

In the world of forensics, facts can sometimes be as mind-boggling as fiction because a similar incident happened in real life. The Finnish police apprehended a potential thief, thanks to a dead mosquito detective. After the police found the stolen car, they looked for any clues left behind that could lead them to the culprit. Disappointed by the lack of evidence, they were about to retreat, but that's when one of the officers spotted a dead mosquito whose stomach was filled with blood and had a hunch: what if the mosquito bit the culprit? So, they sent the mosquito for analysis. They got a hit! The DNA of the blood matched the profile of a man who was already in the police records for other crimes. While the jury was out on whether he was the thief, narrowing down a potential suspect using a dead mosquito is no small feat.

An ally for those seeking the truth behind a crime and a formidable foe for those trying to hide their crimes, forensic science, if and when needed, can identify the metaphorical molehill by scrutinising the mole. Through meticulous examination, forensic science can untangle complex clues, piece together molecular puzzles, and unearth evidence hidden from the naked eye. Forensic science has become an indispensable tool in the evidence-based criminal justice system. But was this always the case?

Is justice blind?

Justice wasn't always blind as we know it to be; the concept that criminal law should be grounded in evidence is fairly new. Even a couple of centuries ago, court verdicts hinged on a jury's decision, often swayed by public opinion about the accused. A high likelihood of being charged loomed over the accused if they belonged to a lower social class, were new to the town, had a different skin colour, or were just in the wrong place at the wrong time.

A prime example of how prejudice against a person could easily take someone to the gallows was the death of Stanislaus Bilansky and the conviction of his wife, Ann Bilansky, who became the first woman ever hanged in the state of Minnesota.

Stanislaus Bilansky was a Polish settler who moved to Minnesota in 1842 and earned his living initially as a tailor but later transitioned into the roles of storekeeper and barkeep. Soon, he became a known drunk in the town whose disposition could be described as mostly sad, with frequent outbursts of anger, which led to abuse of his family. After his previous wife, Ellen, left him and their four children behind, in 1858 he married Ann, who was a widow at that time.

Almost a year after their marriage, in February 1859, Stanislaus fell ill with symptoms such as fever and vomiting. A brief period after the appearance of the symptoms, he succumbed to his illness on 11 March 1859. The coroner's team quickly tested the body and found no evidence of foul play, declaring it a natural death.

The day his body was buried, the police got a call from the Bilansky family's neighbour, Lucinda Kilpatrick. She told the police about her suspicions regarding Ann. She believed that Ann might have had a hand in Stanislaus's death. She recalled Ann buying arsenic a few weeks earlier to kill rats, but now Stanislaus's sudden death led her to believe it was for something more sinister. So, the police dug up his body to test for arsenic. Lo and behold! The autopsy did reveal something similar to arsenic.

'Similar' is the key word here because the doctor who conducted the tests wasn't sure if the inflammation in the stomach which is often considered a sign of arsenic poisoning was actually from arsenic or his drinking habits. Furthermore, only one out of six chemical tests used to detect arsenic came out positive. And yet Ann was arrested and sentenced to death for the murder of her husband.

Why? Because the trial of Ann Bilansky wasn't based on evidence but rather on the testimony of star witness Lucinda Kilpatrick, the neighbour. She claimed in court that Ann had an inappropriate relationship with her own nephew, John Walker. Stanislaus disapproved of their relationship, which led Ann to poison him. She also pointed out in court that Ann didn't look grief-stricken enough at her husband's funeral.

Swayed by this narrative of an unfaithful woman with an inappropriate relationship, created by the neighbour, the jury found Ann guilty despite the evidence against her being ambiguous. Over 160 years after her death, there is a lingering suspicion that the outcome of Ann Bilansky's case would have been different if forensics had been involved. There was no way to challenge the court's decision without the right way of analysing the evidence.

Facts over fame

Once you look past the intimidating laboratory equipment and the tongue-twisting scientific jargon, forensic science, at its core, is the application of science and its methods to examine the evidence and uncover information that can help with the decisions that are made in matters of criminal and civil law. A justice system rooted in science relies on facts, not fame. The objective nature of scientific methodology provides a level playing field for all parties involved, regardless of their social status or popularity. John White Webster, a lecturer of chemistry and geology at Harvard Medical College, and his defence team learned it the hard way.

Webster, a part of the Boston high society, belonged to an established family in the area, the kind of family who lived in a mansion. Apart from his lineage, his profession also brought him into the good books of the society around him. So, when he was accused of murdering another man belonging to his class, it took Boston by storm.

In the autumn of 1849, Dr George Parkman, a renowned doctor, philanthropist and landlord of many properties in

Boston, was on one of his long, solitary walks across the city collecting rents and debts. However, Parkman vanished after this trip and was never seen again. As the report of his being missing was filed, the police jumped into action to find the famous and wealthy doctor.

Soon after the news of Parkman's disappearance spread across the town, some facts came flooding back to Ephraim Littlefield, the janitor who worked at Webster's laboratory. He remembered Webster's lab being locked from the inside on the day Parkman disappeared. Webster was angrier than usual, spending time in the lab way beyond his usual hours, but what stood out the most to Littlefield was Webster gifting him a giant turkey as a Thanksgiving present.

So, over the Thanksgiving break, Littlefield started to look for clues in Webster's personal lab space, and his intuition didn't fail him. After excavating a part of the wall, he was hit by a foul stench. He immediately called the police, who discovered a human pelvis, a thigh and a calf. The cops found a jawbone and teeth in the laboratory furnace.

Webster was immediately arrested on suspicion of murdering Parkman, and his case went to trial. The Boston elites were shaken by this turn of events. They gave out public statements vouching for his character and how he was wrongly arrested. Some even pointed the finger at the janitor for trying to frame Webster, as well as to get the reward money for finding Parkman's body.

During a trial that became an overnight obsession, drawing crowds of thousands, Webster's defence team attempted to use public sentiment to their advantage. They tried spinning the tale and presenting it as a case of a malicious janitor trying to frame an esteemed professor. Since the body was dismembered, proving that the body belonged to Parkman beyond a reasonable doubt (a new addition to the criminal justice system) became a challenging task for the prosecution.

Many witnesses were called to identify the body, and many testified that the body was indeed Parkman's from its height and structure. To add to the pile of evidence were eyewitnesses who saw Parkman enter the building where Webster worked the day he went missing but never saw him leave. Due to the circumstantial nature of the testimonies, the court was still reluctant to make a decision – until one crucial piece of evidence, which in today's world might be called forensic evidence, solidified the case.

Remember the jawbone and teeth found by the police? Nathan Cooley Keep, Parkman's dentist, testified in court that he'd fashioned some false teeth for him a few years back, so if the body belonged to Parkman, the jawbone must have false teeth. And it did!

He explained to the jury that to create false teeth, he had to make a dental cast in plaster. He then went on to demonstrate that the jawbone found perfectly fit the cast, thus convincing the jury that the body found in Webster's lab was indeed Parkman's. This case set a precedent for using a piece of forensic evidence and scientific methodology to bring justice.

Following Keep's demonstration, John White Webster was found guilty of the murder of George Parkman and was sentenced to death. Some people were still not convinced, but to their disappointment, Webster confessed to the murder a few days before he was executed.

You might be wondering, what was the motive behind the murder? It was money. To the outside world, the professor was a wealthy man who lived in a castle in Cambridge, but in reality he was in debt. He borrowed a lot of money from Parkman and couldn't return it. The day he went missing, Parkman visited him to reach an agreement on the money situation. Little did he know that the professor had other plans.

Tracking the trace

The laptop keyboard you touched today, and the cover of this book you are holding right now, retain invisible remnants of your presence because 'every contact leaves a trace'. Coined by Edmond Locard in the seventeenth century, this phrase has been the guiding principle of forensic science. With advances in science and technology, the traces left behind can speak louder than ever, helping the justice system keep biases at bay as much as possible. Does that mean forensic science is foolproof? The answer is no. It comes with its own set of pitfalls. To truly understand these, we must embark on a journey into the depths of forensic science, exploring its triumphs, trials and tribulations.

WHAT IS FORENSIC SCIENCE? 1

Rise of forensics

Argentina's Buenos Aires, on 29 June 1892: Ponciano Caraballo comes home to a gruesome scene in his bedroom. His six-year-old and four-year-old children lay lifeless on the bed with their throats slit open; beside them, his wife, Francisca Rojas de Caraballo, was still alive but had sustained superficial injuries. Since she was an eyewitness, the police began questioning her about what had happened. She pointed the finger towards Pedro Ramón Velázquez, a labourer at a nearby ranch, stating that he had killed her children when she rebuffed his attempts to seduce her and refused to marry him.

Velázquez was immediately arrested and, when he refused to confess, was tortured by the police. They locked him up in a cell overnight with the children's bodies, and a policeman even dressed up as a ghost to scare him and elicit a confession from him, but nothing worked. He stood firm that he was innocent. With no solid leads, the police began exploring other ways to tie him to the crime. Their suspicion also shifted toward Francisca, whose story kept changing with every retelling.

So, investigators returned to the crime scene for another sweep in hope of finding some new clues. They combed through every inch of the room, and there it was – a bloody fingerprint on the door.

The part of the door containing the print was removed and taken to Juan Vucetich, an Argentinian police official, who was working on developing a workable system of fingerprint identification. He compared the bloody fingerprint pattern with those of Velázquez and Francisca, and there was a match, not with the labourer but with the mother.

Francisca Rojas swore she had never touched the bloody bodies, but the bloody print proved otherwise. When faced with the evidence, she broke down and confessed. She thought it was better to kill the children than give them up to her estranged husband.

And that's how the first murder case was solved using fingerprint evidence.

What is forensic science?

In our current understanding, the term 'forensics', often used interchangeably with 'forensic sciences', refers to the application of tried, tested and widely accepted scientific knowledge to interpret evidence. This includes both physical evidence and facts within the context of a criminal or civil case. A majority of forensic work, apart from evidence collection, takes place inside facilities that are isolated from the outside world to ensure safety and prevent contamination.

The word forensics, though always tied to legal procedures, centuries ago evoked a very different mental image of a physical space in people's minds. The term is derived from the Latin word *forensis*, which means of or before the forum.

During the Roman era, a criminal charge was followed by a debate between the accused and the accuser at a public square before a forum of people. Both parties were given the opportunity to deliver a speech or present evidence to defend their stance and persuade the forum. The person with the best forensic presentation was declared the winner. Its meaning gradually shifted from public speeches to the use of standard methods to verify or challenge information gathered during an investigation.

Before lab coats and million-dollar facilities, there were town wise men or jurors who, with their understanding of human nature, tried to catch the bad guy. For most people, telling a lie can create stress and anxiety, especially when interrogated about it. This can reduce saliva production, causing a dry mouth. It is said that judges in ancient India recognised this phenomenon and used it effectively to catch thieves. Suspects were gathered and interrogated about the theft, then asked to fill their mouths with rice, hold it for a while, and spit it out. If the rice came out completely dry, it was taken as evidence of guilt, since nervousness from lying caused the person's mouth to become dry, preventing the rice from becoming moistened.

A second-century instructional note by Roman legal expert Quintilian recounts the case of Paries Palmatus, also known as The Wall of Handprints. A man was found murdered with a sword sticking out of his body. His blind son was accused of murdering his father to get his hands on the inheritance, since he was named the heir. The evidence against the son was his sword lodged in his dead father's body and a trail of bloody handprints on the walls leading to his room. In his defence, the son stated that he was blind, and so navigating to his father's room from a secluded part of the house where he lived in the darkness of the night was next to impossible for him. The son had a feeling that it was his stepmother who murdered his father and framed him to seize the inheritance.

The defence lawyer analysed the evidence at the scene and concluded that it was indeed the wife who killed the old man. Firstly, the wife was sleeping next to him that night, which was suspicious given that she slept through the entire ordeal. Secondly, why would the murderer choose nighttime to kill the father and risk having an eyewitness?

Now, coming to the handprints, they were very regular and evenly spaced, and it was suggested that it would be difficult for a blind person to accomplish this. A hand dragged across the wall would make more sense. Even if the son did press his hands against the wall and leave those marks, the first one would be the darkest due to a hand soaked in blood, and the marks would eventually fade. In this case, the prints seemed to be refreshed after a certain distance, indicating a deliberate attempt to make the handprints. Furthermore, a hand that was clenched against a sword a few moments ago would leave an incomplete print owing to the temporary dent caused by holding the sword. All the evidence pointed towards a sighted person with a motive to frame the young man, in this case the stepmother.

If read without its historical context, this passage could easily be mistaken for a modern forensic bloodstain analysis and crime scene reconstruction.

While these are some obscure or lesser-known examples of the role of forensics in the ancient legal system, the next one has been right in front of us all along. Let me give you a hint: Eureka!

If you guessed this was the work of Archimedes, you are spot on. The legend has it that the Greek mathematician and inventor Archimedes was tasked with figuring out a way to prove that King Hieron II of Syracuse was scammed by the goldsmith who made his crown. The King suspected that his crown wasn't pure gold, but was a mixture of gold with other, cheaper metals. The only condition posed to Archimedes was that the crown was not to be damaged. With his hands tied by this condition, Archimedes busied himself with research. When he felt lost, he decided to immerse himself in a comfortable bath. As he lowered himself into the water, he realised that his body pushed aside a volume equal to the volume of his body. Inspired, he tested King Hieron's crown by comparing the volume of the water it displaced with that of a pure gold sample of equal weight. When the gold displaced less water, Archimedes revealed the goldsmith's deceit – the crown was not pure gold, because it was less dense.

Similarly, experts detecting jewellery and art fraud also test the suspected sample against established standards to assess its legitimacy. For instance, every element, when hit with X-rays, interacts with the waves in its unique way and emits its own waves. These emitted waves produce a unique fingerprint pattern on the detector. Since gold has a well-defined spectral line, experts use X-ray fluorescence to bombard jewellery with X-rays without causing any damage and compare the resulting spectrum with the standard spectrum of gold. This process quickly uncovers hidden metals and reveals the exact caratage of the gold.

Forensic science enters the criminal justice system

In the early eighteenth century, Lord Gilbert presented a law of evidence that stands true even after three centuries. It stated that the law must demand the utmost evidence that the nature of the fact is capable of. In other words, courts should seek the best possible proof.

In the English legal system, experts have been part of trials since the Middle Ages. They were considered a neutral body appointed by the court, but the Adversarial Revolution in the late 1700s changed that. This new ideology stated that the truth emerged from the clash between opposing parties, rather than from a court-controlled investigation. Thus, an expert who became a partisan witness of the disputing parties was chosen and paid for by them.

It was becoming quite evident that, in many cases, the way to obtain the best evidence was to look through the lens of science. As the Industrial Revolution rolled out across the UK and the USA, so did disputes with scientific and technological developments at their core. The shift in the nature of disputes further made scientific expert witnesses a controversial yet indispensable part of the courtroom. By the early nineteenth century, the jury was not only hearing from traditional witnesses but also from chemists, physicians, microscopists and numerous experts in their respective fields.

While the number of scientific witnesses grew, the hard science of forensic lab work still lagged behind. Forensic autopsies were the exception rather than the norm in the eighteenth century and well into the nineteenth century, even in homicide cases, where trials often relied almost entirely on witness testimony, as seen in the conviction of Ann Bilansky.

It was only in the mid-nineteenth century that institutions started providing courses for forensic sciences, with some medical schools in Britain even making them a compulsory module in the medical curriculum.

A similar trend of incorporating forensics into medical courses was also observed in China. In 1915, when the

Ministry of Education of the Republic of China issued the Medical School Rules, medical schools developed a curriculum that included judicial medicine and judicial chemistry. These disciplines were later renamed forensic medicine and forensic chemistry, terms still in use today.

A few years later, between 1929 and 1935, institutions like Zhejiang Specialty Medical School introduced dedicated forensic courses for advanced training. Graduate students from medicine and pharmaceutical science enrolled in these programs and underwent six additional months of intensive study in forensic medicine. Upon completion, they were appointed as forensic specialists in provincial district courts, directly linking academic training to judicial practice.

In tandem with the medico-legal practice, another crucial pillar of forensics was taking shape inside a laboratory in London.

Forensics gains popularity

Sir Francis Galton, born in 1822 near Birmingham, England, was a man of many talents. He was an explorer, polymath, anthropologist and statistician.

His interest in investigating heredity in humans was sparked by the publication of *On the Origin of Species* by his half-cousin Charles Darwin in 1859. Galton's fascination with heredity drew him to anthropometry – the measurement of human traits. Realising that reliable statistics required a large dataset, he set up an Anthropometric Laboratory at the 1885 International Health Exhibition in London, where he gathered vast amounts of data on human characteristics.

Fingerprint impressions were part of the data that Galton gathered. He collected and analysed over 8,000 fingerprint samples and went on to classify the patterns into arches, loops and whorls. He also demonstrated that fingerprints are unique to an individual and the patterns remain consistent throughout a person's life.

He laid the foundation for modern fingerprinting technology in three books: *Finger Prints* (1893), *Blurred Finger Prints* (1893) and *Fingerprint Directory* (1895). His classification system began its journey outside the library and in the criminal justice world after Sir Edward R. Henry, Commissioner of Police of the Metropolis in London, created his own system by combining his research with Galton's.

While the Henry Classification System is often associated with England, its origins are deeply rooted in India. It was developed at the Calcutta Anthropometric Bureau by Rai Bahadur Hem Chandra Bose, an Indian police officer and mathematician, under the leadership of Henry, who was the Inspector General of Police in Bengal. This classification system was first accepted in India, leading to the establishment of the world's first fingerprint bureau in Calcutta (now Kolkata) in 1897, four years before its adoption in England and Wales and the founding of the first United Kingdom Fingerprint Bureau at Scotland Yard, the headquarters of the Metropolitan Police in London. After being adopted by Scotland Yard, it soon spread across the globe.

Galton was sure of his observations but also encouraged further improvement with time, which is reflected in his book *Finger Prints* where he writes:

> Enough was then seen to show that the subject was of real importance, and I resolved to investigate it; all the more so, as the modern processes of photographic printing would enable the evidence of such results as might be arrived at, to be presented to the reader on an enlarged and easily legible form, and in a trustworthy shape. Those that are

> put forward in the following pages admit of considerable extension and improvement, and it is only the fact that an account of them seems useful, which causes me to delay no further before submitting what has thus far been attained to the criticism of others.

Forensic medicine and science were gaining momentum at the dawn of the twentieth century. A sensational case that solidified its role was the trial of the infamous wife-murderer George Joseph Smith, in what the media liked to call the 'Brides in the Bath' case.

George Joseph Smith's rap sheet went back to when he was nine years old, when he was in and out of a reformatory due to theft and fraud-related charges. His crimes related to wives began when he convinced his first wife, Caroline Beatrice Thornhill, a domestic servant, to steal from her employers. They both ended up serving a sentence, and after being released, his wife moved across the Atlantic Ocean to put some distance between herself and her shifty husband.

This didn't stop Smith from pursuing his schemes. Though he wasn't blessed with striking looks, his true power lay in his uncanny ability to charm and manipulate, persuading people to do whatever he desired.

Between 1912 and 1914, three of Smith's wives died in eerily similar ways. Each was found drowned in her bathtub, not long after the wedding. At first glance, their deaths seemed like unfortunate accidents. Bathrooms in those days were small and cramped, while bathtubs were narrow. A fainting spell, a slip or a sudden seizure seemed like possible explanations.

But when the third bride, Margaret Lofty, died in her bath in 1914, suspicions grew. The similarities between the deaths were too unlikely. All the women were newly married, had drawn out fresh insurance policies with their husband as the beneficiary, and had died while alone in the bath. When the case reached the court in 1915, this was the information they had:

1. Bessie Constance Annie Mundy died in 1912. Smith, alias Henry Williams, told relatives she died in a fit in the bath. The doctor called to the scene said death was due to drowning after an epileptic seizure.

   Coroner's Verdict: Death by misadventure from a seizure in the bath.

2. Alice Burnham died the following year in 1913. Smith, alias G. Smith, wrote to her family that she had pains in the head, suffered a seizure, and drowned due to a weak heart. Dr George Billing ruled the cause of death was drowning, likely triggered by a faint from heart disease and hot water.

   Coroner's Verdict: Accidental death from a fit or faint in the bath.

3. Margaret Elizabeth Lofty, the third victim, died in 1914. This time, as John Lloyd, Smith claimed that he found her collapsed in the bath after complaining of dizziness and headaches. The doctor who examined the scene concluded suffocation from drowning, possibly linked to influenza and fainting.

   Coroner's Verdict: Accidental drowning in the bath.

The police grew suspicious of this string of similar types of death. They started piecing the puzzle together, linking the three deaths and tracing them back to the same man: George Joseph Smith.

Sir Bernard Spilsbury, a British pathologist hired as the expert medical witness for the case, began reviewing all the available medical evidence. He examined Bessie Mundy's body after its exhumation on 18 February 1915, with the intention of studying the brain and heart to cross-examine the initial claims of seizures, but severe decomposition prevented him from doing so. He also reviewed the post-mortem examinations in the cases of Alice Burnham and Margaret Lofty, and it was then that he began to notice patterns. As a result, he began running an experiment recreating possibilities of drowning with the help of an expert female swimmer of a similar build to the brides.

Spilsbury testified on the unlikelihood of accidental drowning in the specific bath conditions and body positions described in the cases. He stated that it was almost certainly impossible for a person's head to become sufficiently submerged to cause death by drowning if they were sitting at the broad end of the bath and fell backward during a faint or fit. He had each of the actual bathtubs brought into court, including the one in which Smith's last bride had died.

With the help of his assistants, Spilsbury began the demonstration. He asked his assistant to relax in the bath, just as the women would have, and then suddenly he grabbed her ankles and yanked them upward. This would cause the woman's head to slip beneath the water in an instant. The shock of being plunged underwater, combined with the position of the body, would cause her to lose consciousness almost immediately. Drowning followed within seconds of this effortless yet chilling move. This explanation aligned with both the post-mortem and crime scene evidence.

In August 1915, George Joseph Smith was hanged at Maidstone Prison. Spilsbury's expertise in forensic pathology and the demonstration convinced both court and country how careful reconstruction and expert testimony could unmask crimes masquerading as accidents.

The Metropolitan Police sets up a dedicated crime laboratory

> How science is being summoned to aid in the detection of crime is vividly illustrated by the opening of London's first police laboratory at Hendon ... Over the years, brilliant Home Office analysts, including the renowned Sir Bernard Spilsbury, assisted on major cases. Now, however, Hendon marks a turning point: a dedicated laboratory for criminology under the direction of Dr. James Davidson of Edinburgh University, with departments in toxicology, photography, and advanced techniques already pioneered in America and across Europe.

This excerpt comes from an article titled 'The Scientific Detective: Chemistry of Crime', published in *The Argus*, an Australian daily newspaper, reporting on the landmark opening of London's first police forensic laboratory – the Metropolitan Police Laboratory (MetLab) at Hendon on 10 April 1935. Britain's first dedicated police forensic lab opened its doors with just six employees: a pathologist, a chemist, a police liaison officer, a technician, a clerk and a cleaner. This small team was equipped to handle toxicology, ballistics, and crime scene photography and identification, thus marking the beginning of a long-standing partnership between science and law enforcement in the fight against crime.

In 1872, when India was still under British colonial rule, the Indian Evidence Act was introduced. Among its many provisions, Section 45 specifically recognised the importance of expert opinions in judicial matters. It stated: 'When the Court has to form an opinion upon a point of foreign law or science, or art, or as to identity of handwriting [or finger impressions], the opinions upon that point of persons specially skilled in such foreign law, science or art, [or in questions as to identity of handwriting] [or finger impressions] are relevant facts.'

After independence, India witnessed a remarkable surge in the field of forensic science. A landmark moment occurred in 1952, when Calcutta, home to the world's first fingerprint bureau, also became the site of the country's first State Forensic Science Laboratory. This wasn't just the birth of a lab; it was the beginning of a new era, bringing scientific investigation closer to the justice system.

A fresh fingerprint on paint and the rise of forensics in the US

In the early hours of 19 September 1910 in Chicago, Mary Hiller woke up from her sleep as she heard some faint disturbance inside the house. Her suspicions grew as the gaslight at the top of their stairway, which she always left burning overnight, had gone out. She asked her husband to investigate. As Clarence moved through the house, he came face-to-face with an intruder. A brief struggle followed, ending with several gunshots that left Clarence Hiller dead.

The intruder, Thomas Jennings, a recently paroled convict, couldn't even cross a mile before he was captured with a .38 revolver and a bloodied coat. When questioned, he denied his involvement with the burglary and murder, claiming that his bloody state was due to injuries incurred after falling from a trolley. The investigators didn't believe his story.

After a detailed search of the crime scene, the police discovered a clear fingerprint impression in fresh paint on a railing inside the Hiller home. They removed the porch railing, took an impression of it as evidence, photographed it, and enlarged it. Using Galton's system, they cross-referenced Jennings' fingerprints from his prison records with the one left in paint, and it was a match.

During the trial, the enlarged images of both sets of fingerprints were presented for the jury to compare with the prison records of Jennings's fingerprints. An impressive lineup of experts reinforced the evidence. Two seasoned Chicago police officers took the stand, joined by an inspector from Canada's Dominion Police in Ottawa and a US government investigator who had been trained at Scotland Yard. They unanimously agreed that the prints on the railing were indeed Jennings's.

Jennings's defence team attempted to undermine the credibility of fingerprint evidence, arguing that the technique was new and not well understood. To prove their point, the defence even collected fingerprints from the public, hoping to find a duplicate that would disprove the principle of uniqueness. However, this move quickly backfired, making the prosecution's case even stronger.

Convinced by what they saw, the jury returned a unanimous guilty verdict, and Jennings was sentenced to death. This made the case People vs Jennings the first ever murder conviction in the US based on fingerprint evidence.

Almost a decade and a half later, in 1923, the Los Angeles Police Department became the first in the US to establish a forensic lab. Officer Rex Welsh was appointed as the department's first criminalist by the LAPD Chief August Vollmer. The first publicly funded crime laboratory was a small space at the Central Police Station, where Welsh worked, handling forensic work with just a microscope, basic chemicals and glassware.

A few years later, Ray Pinker, the first civilian police forensic scientist, joined the lab. With a pharmacology degree from USC, Pinker elevated the lab's forensic capabilities, bringing scientific rigour and credibility with innovative methods. Despite these modest beginnings, the crime lab marked the dawn of modern forensic police work in America and became a model for other forensic laboratories to follow.

If you are someone who watches a lot of crime shows, you must be wondering why the FBI hasn't made an appearance yet. The FBI was a bit late, but it eventually joined the party in 1932 with the establishment of the Criminological Laboratory, thanks to its first director, J. Edgar Hoover's insistence on applying scientific principles to aid in criminal investigations. One of the first cases they solved was the case of the deadly chocolate box.

In November 1933, a mental patient named Harrington Fitzgerald Jr. in a Pennsylvania veterans' hospital received a box of chocolates from someone identified only as 'Bertha'. Assuming it was an early Christmas present, he indulged in a significant portion of the chocolate. Soon, Fitzgerald fell gravely ill, and died soon after. Since the crime occurred on federal property, the Department of Justice took over the investigation.

Primary investigations pointed towards the victim's sister, Sarah Hobart. The only evidence that could link this case to her was the handwriting accompanying the box. So, the agents gathered handwriting samples linked to her, along with the chocolate wrapper and the accompanying card, and sent them to the Bureau's fledgling laboratory for analysis.

At the lab, Special Agent Charles Appel, the Bureau's sole forensic analyst at the time, compared the writing on the note from Bertha to Hobart's handwriting samples. He didn't find a conclusive match. Appel then suggested the investigators locate Hobart's typewriter, but they couldn't find it at her place, for she had sent it for repair. The agents tracked down the typewriter and compared the typeface with the print samples on the mailing table – they found a match! With the evidence stacked against her, Sarah couldn't hide behind her lies anymore, so she ended up confessing to the crime.

SHERLOCK AND SCIENCE 2

> 'Dr Watson, Mr Sherlock Holmes,' said Stamford, introducing us.
>
> 'How are you?' he said cordially, gripping my hand with a strength for which I should hardly have given him credit. 'You have been in Afghanistan, I perceive.'
>
> 'How on earth did you know that?' I asked in astonishment.

This excerpt from *A Study in Scarlet* marked the first ever interaction between Sir Arthur Conan Doyle's legendary detective, Sherlock Holmes, and Dr John Watson, a retired army doctor who would go on to become Holmes's friend, flatmate and biographer. Watson was taken aback when Holmes deduced his return from Afghanistan, because how could that be possible when they had just met for the first time? Unless someone had got Holmes up to speed about his past.

This assumption Sherlock brilliantly refuted later on in the novel: 'I knew you came from Afghanistan. From long habit the train of thoughts ran so swiftly through my mind that I arrived at the conclusion without being conscious of intermediate steps. There were such steps, however.'

As soon as Watson entered, Sherlock noted that his face was tanned, indicating he had recently returned from a region with a hot climate and ample sunlight. He observed that Watson's left arm was held in a stiff and unnatural way, suggesting an injury. So, by combining the tan, the injured arm, and the knowledge that Watson was a military doctor, Holmes concluded that Watson had served in Afghanistan and sustained a wound there.

Readers first witness Sherlock Holmes' iconic investigation techniques in the chapter titled 'The Science of Deduction'. He uses a technique called abductive reasoning, a way of thinking that looks at the clues in front of you and asks: what is the most likely explanation for these facts?

Instead of just listing what is known, it builds a hypothesis, a possible story that ties the evidence together and makes sense of what happened. Can this be applied to real criminal investigations? While deduction does have its place in modern criminal investigation, forensic science relies on scientific analysis where the results lead to conclusions. Sherlock Holmes' reasoning often relies on backward thinking – a method where the detective starts with what is already present, such as a death, an injury, or unusual behaviour, then asks himself, 'What must have happened for this to occur?' Then, using keen observation, sharp logic and the power of intuition, the detective reconstructs events step by step in reverse. This process echoes Holmes' famous statement, 'When you have eliminated the impossible, whatever remains, however improbable, must be the truth.'

The process of eliminating everything impossible makes for an impactful dialogue for a story, but when put to the test in real life, several holes start to appear. Mapping out the possible train of events is extremely challenging, as it requires prior knowledge of all variables that could influence the outcome.

Let's break it down with two simple examples:

1. At the start of *The Adventure of the Blue Carbuncle*, Holmes is presented with Henry Baker's forgotten old felt hat, found along with a Christmas goose. Without even meeting the owner, he deduces that it belongs to a middle-aged person who was once wealthy.

Sherlock's reasoning: the owner was wealthy three years ago, enough to afford an expensive hat, but has since fallen on hard times, as he has not replaced it since then.

Other possibilities: this hat is his favourite, or it holds some sentimental value to him, so he prefers to wear it wherever he goes. Or maybe, the man was never rich; some wealthy relative of his must have given him the hat as a Christmas gift.

Then, going back to Holmes' and Watson's first interaction:

2. Holmes accurately guesses Dr Watson's profession, type of injury, and where he was posted last. No matter how logical or well thought out it might be, the deduction was a lucky guess. Why?

By the 1880s, when *A Study in Scarlet* was written, the British had colonised vast expanses of land, including regions in the Indian subcontinent and numerous African colonies. So Watson could have been stationed in any of these colonies, and most of these places would have given him a tan. Afghanistan was not the only option.

Holmes detected what he described as 'the air of a military man', and concluded that the individual must have been an army doctor. But what if he had served in the Navy instead? Or perhaps he wasn't in the military at all, but had been raised in a military household. It's not uncommon for children of service members to adopt the habits and demeanor typically associated with military life.

These 'deductions' sit perfectly within the constraint of Doyle's plotline. It is not impossible, but rather impractical, as one would have to sit and chalk out all the possible outcomes.

Sherlockian science

Apart from using his brain to solve crimes, Sherlock was also a believer in using science to solve crimes, which is quite evident from his array of laboratory equipment in his abode, his chemical analysis of types of ashes and his demonstration of the most practical medico-legal discovery for years, an infallible test for bloodstains.

In *A Study in Scarlet*, Sherlock Holmes demonstrates a chemical process to detect haemoglobin. He begins by pricking his finger with a pin and drawing a drop of blood into a chemical pipette. This drop is added to a litre of water, creating a solution with a blood concentration of about one part in a million. Though the mixture appears clear, he adds a few white crystals, followed by several drops of a colourless fluid. The reaction mixture turns a dull mahogany colour, and a brownish dust precipitates at the bottom. Holmes goes on to explain that the reagent he added precipitates haemoglobin and nothing else, making it a highly sensitive and specific test for identifying bloodstains, whether fresh or old.

A modern equivalent of this would be a presumptive test, such as the Kastle–Meyer Test where the reagents change colour in the presence of blood. The test relies on the iron present in the haemoglobin of red blood cells. The iron and hydrogen peroxide oxidise phenophthalin (a colourless chemical indicator in the test solution) into phenolphthalein (the pink form of the indicator).

This highly sensitive test can detect blood in solution at dilutions of up to 1:500,000, while a visible pink colour is typically observed at dilutions of up to 1:10,000 in the case of blood on cloth.

Back to Sherlock. He also reveals his enthusiasm for chemical analyses with a study of over 140 types of tobacco ash – a fact made clear in his exchange with Watson in *The Sign of the Four*, where he states, 'To the trained eye there is as much difference between the black ash of a Trichinopoly and the white fluff of bird's-eye as there is between a cabbage and a potato.'

A sentiment modern forensic fire analysts would recognise is that ash colour is a critical forensic indicator in fire investigations, offering insights into combustion temperatures, materials involved, and the presence of accelerants. While interpretation can be complex, trained investigators use ash colour alongside other physical evidence to reconstruct fire events and assess intent or cause.

Patterns and distribution of ash colours can help map fire progression and identify points of origin. Heavier, whiter ash may indicate intense, prolonged exposure to heat – potentially pointing to where a fire started. Combining ash colour observation with laboratory analysis (such as mass spectrometry or chromatography) can identify residual chemicals, thus offering evidence of deliberate arson or accidental causes.

On 16 October 2005, at 11am, Sherlock Holmes – once described as a first-class chemist by Young Stamford in his introduction to Watson – was awarded an Extraordinary Honorary Fellowship by the Royal Society of Chemistry to celebrate his love for chemistry and for being the first detective to exploit chemical science as a means of detection. The ceremony was held at the Sherlock statue outside Baker Street Station, to commemorate the hundredth anniversary of *The Hound of the Baskervilles* being published and the knighthood of the character's creator, Sir Arthur Conan Doyle.

So, we return to our sub-titular question: was Sherlock Holmes the first forensic scientist? Perhaps not, even though he showed a keen interest in medico-legal procedures, as his calibre as a detective was defined by the use of logic and thinking backward. His scientific curiosity seemed more personal than professional, and not strictly aligned with forensic science as we understand it today.

Yet, Holmes *did* inspire one of the real pioneers of forensic science. But before we meet him wouldn't you like to meet the man who inspired Sir Arthur Conan Doyle's legendary detective?

The real Sherlock Holmes?

With fingertips pressed together and eyes locked on the man just outside the door, the wiry figure in the long coat leaned back in his chair. His medical assistant asked to let the man in, as he didn't look very comfortable out in the cold. The man in the long coat asked his assistant to bring the man over to the front of the room so they could take a better look at him.

As the man outside stepped into the room, the wiry figure in the chair leaned forward slightly and turned to his assistant. 'Tell me,' he said calmly, 'what troubles this man?'

The assistant blinked, caught off guard. How could he possibly know something troubled him? He had never seen the man before, let alone spoken to him. There had been no conversation, no clues offered – only a tired face and a hesitant gait. All he could offer, somewhat uncertainly, was that the man looked anxious and worried. But then again, they almost always did.

The man in the chair, however, wasn't convinced. He pointed out that the clues were already there. Everyone looks, but few pay attention to what actually matters.

He then went on to say that the man who walked in was suffering from a severe back ache as a result of carrying bricks all day.

The hod carrier and the man's assistant both gave shocked and impressed looks, mystified as to how he had diagnosed this ailment.

So, who do you think the man in the chair was? Did you imagine this scene unfolding at 221B Baker Street, where Doyle's creation was once again baffling Dr Watson with his razor-sharp observations?

In fact, this scene played out at the clinical surgery amphitheatre at the Royal Infirmary of Edinburgh. And while Sherlock Holmes was nowhere to be found, his creator was very much present. Seated among a group of captivated medical students, Arthur Conan Doyle furiously scribbled notes, his mind racing to keep up. At the centre of it all stood Dr Joseph Bell, Doyle's brilliant mentor at medical school. It was here, watching this extraordinary man unravel medical puzzles with logical deduction, that Doyle found the inspiration for his greatest creation.

To quote Ely Liebow, author of *Dr. Joe Bell: Model for Sherlock Holmes*, 'Without Sherlock Holmes few indeed would ever have heard of Joseph Bell; without Dr. Joseph Bell, there would be no Sherlock Holmes as we know him. Oh, what a thing is irony.'

Dr Joseph Bell, born in Edinburgh in 1837, was an excellent surgeon who went on to serve as Queen Victoria's personal surgeon whenever she visited Scotland. He also remained the editor of the *Edinburgh Medical Journal* for decades. Legend also has it that Dr Joseph Bell was consulted by the police in an attempt to help identify the murderer in the infamous and still unsolved Jack the Ripper case. However, there is no firm historical evidence such as Scotland Yard records, case files or contemporary press coverage of Dr Bell's involvement with the case.

During his time as a senior surgeon at the Royal Infirmary in Edinburgh, Dr Bell held captivating diagnostic sessions every Friday at an outpatient clinic, lectures so intriguing that Arthur Conan Doyle and his fellow medical students paid to attend them. The incident described earlier took place during one of these demonstrations. Bell believed the key to being a good physician was keen observation, a principle he drilled into his students week after week. With uncanny precision, he would astonish both patients and onlookers by identifying not only the patient's medical condition, but often their profession, hometown and even their mode of travel to the clinic, all without asking a single question. Beyond the spectacle, Bell emphasised the subtle art of noticing minute details and taught his students the art of perceiving what others overlooked.

If this wasn't enough evidence that Doyle's mentor heavily inspired Sherlock's ability and mannerisms, he even mentioned it in a letter to Joseph Bell, dated 7 May 1892, in which he said, 'It is most certainly to you that I owe Sherlock Holmes.' Some experts believe that a man who frequently appeared alongside Bell as a fellow expert witness in court may have also helped inspire Doyle's creation of Sherlock Holmes.

Littlejohn investigates

On 10 August 1893, Ardlamont House in Argyll, Scotland, was put on every Victorian Brit's map when twenty-year-old

Windsor Dudley Cecil Hambrough, heir to a considerable fortune, accidentally shot himself in the woods.

Early in the morning of the fateful day, Cecil Hambrough, his tutor Alfred Monson, and a third, mysterious figure, Mr Scott (who later disappeared), went on a rabbit hunting trip into the woods nearby. Only two of them came back at 9am; there was no sign of Hambrough or his guns. Monson then informed Wright, the butler of the house, that Hambrough had shot himself in the head and was dead.

The pursuit of the truth about what actually happened in the woods became one of the most sensational criminal trials of that era, and through this, we meet Dr Henry Littlejohn, President of the Royal College of Surgeons, forensic scientist and the Crown's Medical Officer.

Littlejohn was among the first to get a call from the officers when investigating serious incidents. He was regularly called upon to carry out post-mortems and present his findings to the jury during trials. In the Ardlamont case, his testimony in court was scheduled on the third day of the trial, which focused on the nature of Cecil's injuries, the scene of the shooting and the trajectory of the fatal shot.

Littlejohn was tasked with exhuming Cecil Hambrough's body from its grave and conducting a post-mortem examination. He described the fatal wound as being behind the right ear, about three-and-a-half inches in length, triangular, with its base towards the face and apex below and slightly in front of the occipital protuberance. He observed no blackening or singeing around the wound, which would be expected from close-range gunpowder discharge. Littlejohn collaborated with Dr Bell and gunsmith James MacNaughton to carry out crime recreation experiments where they fired shots into cardboard targets and human cadavers at the Edinburgh mortuary to determine firing patterns and their effects.

Littlejohn firmly argued that the gunshot wound could not have been self-inflicted based on the wound and gunpowder discharge pattern. All eyes were on Bell as he took the stand, his testimony highly anticipated following his recent rise to fame as the inspiration for Sherlock Holmes. Bell backed his colleagues' conclusions without hesitation, reinforcing the belief that this was no act of self-harm.

A similar strategy to refute a claim based on gunpowder residue also appeared in a Holmes case, 'The Adventure of the Reigate Squire'. The victim is William Kirwan, a coachman who is murdered while working for the Cunninghams, a wealthy local family. When questioned, the Cunnighams claim that a burglar had entered their estate, and Kirwan tried to stop him, and during the altercation he was shot by the burglar. Upon investigating the body, Holmes notices that William's body has no powder burns on it, so he was not shot at point-blank range as the Cunninghams claimed.

Doyle had the rare privilege of being mentored by those two brilliant minds of his time, Bell and Littlejohn. Drawing on their sharp observational skills and forensic expertise, Doyle crafted the legendary character of Sherlock Holmes. Over a century later, Holmes has appeared on screen more than 200 times, securing his place as one of the most portrayed and endeared figures in film and television history.

Sherlock inspires the birth of forensic practices

Most people show their admiration for their favourite fictional character or actor by either dressing up like them or collecting memorabilia from their work, but not many people in this world would sit down and write an entire research paper to emulate their inspiration.

A young Edmond Locard devoured the works of Sir Arthur Conan Doyle, finding in them both entertainment and intrigue. Something that especially struck him was when Sherlock's peculiar expertise was mentioned in *A Study in Scarlet*: 'I have made a special study of cigar ashes – in fact, I have written a monograph upon the subject. I flatter myself that I can distinguish at a glance the ash of any known brand, either of cigar or of tobacco.' Four decades after the story was published, in 1929, Locard turned Sherlock Holmes' fictional boast into scientific reality by publishing a paper called 'The Analysis of Dust Traces', which detailed the identification and analysis of tobacco ashes found at crime scenes.

His connection with the pipe-smoking detective doesn't end there. Just like Sherlock, Edmond Locard was a detective too – solving crimes from a two-room space on the second floor of a courthouse in France, equipped with nothing more than a microscope and a spectroscope.

This unassuming space – the Laboratoire Interregional de Police Technique, or Interregional Technical Police Laboratory, in Lyon – was the world's first police crime lab, established in 1910. Here, Locard broke from the tradition of detectives mostly relying on intuition and interrogation. It became a place where scientific techniques and legal knowledge converged, allowing cases to be approached through careful evidence collection and systematic analysis, ensuring that logical reasoning guided the path to the truth.

Locard earned his doctorate in medicine in 1902, after which he began working with Alexandre Lacassagne, gaining a deeper understanding of the intersection of medicine and forensics. After his time as Dr Lacassagne's assistant, Locard decided to study law, and in 1907 he successfully passed the bar exam. Now a medical doctor and lawyer, Locard's interest in medico-legal sciences grew even deeper. To get a non-medical perspective on forensics, Locard decided his next stop would be to spend some time with anthropologist Alphonse Bertillon.

Bertillon was working on a criminal identification system called anthropometry, which involved the systematic measurement of human physical features, such as head length and arm span, to distinguish and identify individuals. During this time, Locard also encountered the technique of dactylography, the study of fingerprints used to identify both suspects and victims. Both Henry Faulds, a Scottish doctor, and Sir William James Herschel, a British officer in colonial India, recognised the potential of dactylography, but as we have seen it was Francis Galton who systematically studied it and perfected the technique of fingerprint analysis for identification.

At his crime lab in Lyon, Locard dove deep into Galton's work, and soon enough, an opportunity arose for him to test in the field.

A burglary had taken place at an apartment in Lyon, where the thieves had emptied all the jewellery stored in a rosewood box. There were no eyewitnesses and no visible evidence was left behind by the criminals. The police, however, had two suspects, the infamous burglary duo Boudet and Simonin, and yet they couldn't be arrested since there was nothing that could connect them with the crime scene. However, when Locard was called to examine the crime scene, he found that the rosewood box was covered with numerous fingerprints.

Using his trusty comparison microscope, Locard inspected the pore locations (which vary from person to person) of the fingerprints found on the box and compared them with the suspect's fingerprints. The patterns observed were similar, which meant it was highly improbable that the fingerprints belonged to anyone else apart from Boudet and Simonin. This marked one of the first high-profile cases to be solved using fingerprint analysis.

While Locard's forensic escapades made him a household name in France, what brought him global recognition was his Exchange Principle, a concept that remains the foundation of many forensic practices today.

Locard's Exchange Principle

Between the years 1931 and 1935, Edmond Locard published seven textbooks on criminalistics and the use of forensics for solving criminal cases. Something that is still considered a guiding principle in the field of trace evidence forensics is his (translated) statement: 'Any human action, and even more so the violent action of a crime, cannot take place without leaving some mark. The remarkable thing is the variety of these marks. Sometimes they will be prints, sometimes simple traces, sometimes stains.' This sentiment has been condensed into the popular phrase 'Every contact leaves a trace'.

In essence, Locard's Exchange Principle states that when two things come into contact, there is an exchange of material between them. Even the briefest interaction can leave behind or pick up tiny pieces of evidence that serve as silent witnesses to the event. For instance, when you touch a coffee mug, you leave behind a fingerprint created by the oil secreted by the pores on your fingertips, and in return, you remove some dust particles from your surroundings that had settled on the mug.

In crime scenes, this exchange could leave behind fingerprints, hair, cloth fibres, soil, a button, and even DNA. On the other hand, the suspect can also unknowingly take away soil from the garden, a piece of curtain, the victim's hair, or petals of a very unique flower that only grows in the area near the crime scene.

Locard demonstrated the reliability of his principle during the investigation of Frenchwoman Marie Latelle's death.

Marie Latelle was found murdered by strangulation at her parents' home in Lyon. The suspicion immediately fell on her boyfriend, Emile Gourbin, a bank clerk. When interrogated about his whereabouts at the time of the murder, which the medical examiner said was around midnight, Emile had an airtight alibi. He spent the entire evening with his friends eating, drinking and playing cards, and called it a night only after 1am. His friends verified his claims and vouched that they were with him at the time.

Locard was called to help with the case. His first move was to remove scrapings from under his fingernails because if he was involved in the strangulation, he must have carried back some trace of the incident with him. He studied the scrapings under the microscope and found some tissue which could be Marie's, but he had no way to prove it. He also found a pink dust which he identified primarily as rice starch. Chemical analysis revealed that the pink dust also contained a mixture of bismuth, magnesium stearate, zinc oxide and a reddish iron oxide pigment called Venetian red, often used in women's cosmetics. He took his findings to different druggists in Lyon. One of them recalled that he indeed made a powder for Marie with the same composition. He then asked the police to bring him Marie's makeup, and there it was, a pink face powder in Marie's possession, which was the same as the one found under Emile's fingernails.

When confronted with the findings, Emile confessed that he was furious that Marie had refused to marry him. So he devised a plan where he got his friends extremely drunk, then moved the clock ahead to make them believe it was 1am, knowing they would be too hungover the next morning to notice the actual time at their own homes. When his friends left his home, which was before midnight, he went to Marie's place and strangled her to death.

This case cemented the effectiveness of Locard's Exchange Principle, a legacy that is still evident in modern forensic practices.

EVERY CONTACT LEAVES A TRACE

3

Fingerprints and DNA

On Thursday, 8 August 1963, the Royal Mail train left Glasgow for Euston Station in London. Transporting mail and parcels, the train was also carrying registered mail in its extra-secure high-value carrier, mostly containing cash. On a usual day, the carriage carried items worth approximately £300,000, but due to a Scottish bank holiday, this particular train was carrying a total of £2.3 million, eight times its usual amount.

A few moments after 3am, the train had to come to a halt near Sears Crossing at a red signal. The co-driver got down to check what was wrong, and that's when he was ambushed by masked men and thrown down the steep railway embankment. In the meantime, a masked member of the group entered the driver's cabin and hit the driver on the head to render him unconscious.

One of the robbers had learnt how to drive a train by befriending a railway staff member on the pretext of being a rail enthusiast. While he drove the train a mile from the scene of the ambush to where their getaway cars were positioned, his accomplices held the Post Office staff hostage as they broke into the high-value carriage, with 120 sacks containing two-and-a-half tons of money.

The Post Office staff in other carriages had no idea what was going on in the first two carriages. The robbers left and instructed the staff of the looted carriages to stay put for half an hour before contacting the police: otherwise, there would be consequences. Within minutes, one of the greatest robberies, planned with military precision, had been carried out.

As soon as the news broke, a massive manhunt began, with the Flying Squad, Buckinghamshire Police and the British Transport Police throwing everything they had into the investigation.

The gang had taken shelter in an old farmhouse. A nearby resident grew suspicious of the movement of the people living on the farm and reported them to the police. Upon arrival, the police couldn't catch anyone red-handed, but they found large amounts of abandoned food, provisions and a monopoly board, which became a crucial piece of information as the forensic team found several fingerprints on the game. The fingerprints, alongside a web of clues untangled during the investigation, led to the arrest of each suspect, one after the other.

What is a fingerprint? How are they formed and why are they unique?

Humans have their unique identifiers ready to go even before they are born. Inside a mother's womb, around the tenth to eleventh week, the soft pads on the fingers and palms begin to shrink and fold in unique ways that are dictated by genetic and environmental factors. Over the next few weeks, the folds expand outwards starting from three different points and turn into a complex pattern of grooves and ridges whose twists and turns are shaped by signalling between proteins found in genes.

The primary biological function is to increase friction between the skin and surfaces we touch, helping us grasp things, even those with smooth surfaces. Then, some people discovered that these fingerprints can also help us catch criminals. The first use of fingerprints, as evidenced in Francisca Rojas' case discussed in Chapter 1, changed the course of the entire case and the future of criminal proceedings.

If you recall from that case, the door panel with the bloody print was sent to Juan Vucetich, the Argentinian police official developing fingerprint identification. He matched the print to Francisca, not the original suspect – cracking the case and sending her to jail.

Our focus in this chapter is going to be on what helped him match the prints and led him to the conclusion. Spotting differences in fingerprints is definitely more difficult than a *Where's Wally?* puzzle, but it's not impossible if one knows where to focus.

Juan Vucetich's method was inspired by none other than Sir Francis Galton, the pioneer who had delved deep into the unique patterns of human fingerprints for very different reasons but ended up making a helpful discovery.

Centuries before detectives were scanning crime scenes for these whorly skin impressions, fingerprints were being used as a personal identifier. Their use as signatures for contracts and as proof of identification dates back to 300 BC in China. In Japan, divorce documents bore marks of fingers, used as formal proof of consent. A paper published in *Nature* in 1896 suggests that this tradition may have its roots in China's ancient legal code, probably adopted from the 'Laws of Yung-Hwui' (*circa* AD 650–55), issued under the reign of the third emperor of the Tang, Gaozong.

Fingerprint identification had a modern revival with Sir William Herschel, Chief Magistrate of the Hooghly District in India, when it was a British colony. Herschel required every contract he made with the locals regarding road metalling work and tenders for the supply of construction material to be signed with an impression of palm prints, and later, simply the prints of the right index and middle fingers.

As more and more people started adding finger impressions to sign a binding contract, and his collection grew, he began to notice that these impressions were apparently unique and could help prove or disprove identity.

On the other side of the world in Birmingham in the United Kingdom, Francis Galton was also developing an interest in friction ridges on fingers; in simple terms, fingerprints as an aid in determining heredity and racial background. He was a eugenicist who believed in improving the genetic quality of the human population by inhibiting the fertility of those considered inferior and promoting that of those considered superior. As a result, he amassed over 8,000 sets of fingerprints from anthropological laboratories and from William Herschel's personal collection.

Francis soon realised that while fingerprints didn't offer any clues about a person's intelligence or race, they did provide a unique identity, as no two fingerprints were alike. Even identical twins do not share fingerprints.

He developed an efficient system for capturing inked fingerprint impressions, where he classified fingerprint ridges into three main types: loops, which curve back on themselves; whorls, which spiral or circle; and arches, which rise and fall without forming loops or circles. These patterns then go on to create a tapestry of intricate and distinguishable ridge patterns called minutiae, which can give rise to lakes or enclosures, bifurcations, trifurcations, and ridge endings, among other subtle traits that make every fingerprint truly one of a kind. These identifiable features provide crucial anchor points for forensic analysts and fingerprint recognition software to match fingerprint evidence with records stored in biometric databases.

But are these ridge impressions truly one of a kind?

According to Francis' calculations, the odds of two individual fingerprints being the same were one in 64 billion. While forensic experts believe that the probability of two individuals having a similar set of fingerprints might be extremely low, this estimation of uniqueness might be overly optimistic.

Even the notion that 'no two fingerprints are the same' was challenged by a 2024 study, which used artificial intelligence (AI) to analyse 60,000 fingerprints. With 99.99 per cent confidence, the study found that the swirly parts at the centre of fingerprints from different fingers of the same person share very strong similarities. Does this render fingerprinting an obsolete tool? In fact, it does the opposite.

Imagine a scenario where detectives are solving a serial robbery. In one scene, they found a thumb and a pinky impression of the right hand, and in another, they found a right-hand index fingerprint. Now, the investigations have led them to a suspect, but unfortunately, they only have a left-hand impression of this person, which makes the fingerprint evidence absolutely useless, since fingerprint biometric systems assume that no two fingerprints are identical, even from the same person. However, the researchers believe that, if trained well with even broader datasets, their AI fingerprint detection tool might be able to identify whether all the fingerprints belonged to the same person or not.

How are fingerprints detected and lifted?

The fingerprints we leave behind differ depending on the surface, so crime scene investigators (CSIs) upon arriving at the scene keep an eye out for three types of prints: patent, plastic and latent.

Patent fingerprints are the visible ones that the naked eye can clearly pick up. No special techniques are required to lift such prints, just high-resolution photography is enough to preserve them as evidence. Examples of patent prints would be a paint-stain fingerprint left by a burglar on the window of the freshly painted house he robbed, or a mud stain left on the wall. Patent prints can also be left in ink or even blood, as in the Rojas case.

Patent prints can be formed on surfaces ranging from rough (such as a wall) to smooth (like a tabletop), porous (like cloth) and non-porous (like metal). Apart from a keen eye, spotting a patent print also requires good lighting conditions, so CSIs always make sure to carry a powerful light source or alternate light source such as UV lamp when hunting for patent fingerprints.

The next type is plastic prints, which are three-dimensional impressions left on soft surfaces, like wet paint, wax or soap. When a finger comes into contact with a pliable surface, it tends to leave behind an impression of every curl and whorl of the skin's raised friction ridges; it's almost as if the fingerprint is sculpted into the surface. To spot such prints, forensic experts use special angled lighting. If they believe the print is important and clear enough to be used as evidence, it may be 3D scanned or lifted from the soft surface using a cast made of silicone, sealing wax or dental putty.

Now come the sneaky ones that are both a nightmare and a blessing for forensic experts. Unlike patent and plastic prints, latent prints are invisible to the naked eye and require special forensic pixie dust and potions for them to appear, which makes the task difficult for the CSI but also prevents criminals from finding and trying to remove them. The latent fingerprints are held in natural oils and sweat secreted from the skin which leave a residue on any surface that is touched.

The most common method for making latent prints visible, also made iconic by CSI shows, is dusting surfaces with fingerprint powder. CSIs use their trusty brush and a palette of fingerprint powders to visualise prints from smooth and textured surfaces. The fingerprint-dusting powders stick to oils and moisture left by the ridges on our fingertips to create a visible print, which can be lifted off using transparent tape.

Fingerprint powders usually contain two main ingredients: a pigment, often carbon or metal powders, to bring in the necessary contrast for the print to pop; and a binder like gum arabic or rosin for the powder to stick to the surface. The forensic team has a variety of fingerprint powders at their disposal, ready to dust all the different types of surfaces they may encounter in the crime scene. For instance, aluminium flake powder works best on smooth glass surfaces, but for shiny silvery surfaces, brass flake or magnetic jet-black powder is used to create a better contrast. Fluorescent powders are also used to enhance prints on multi-coloured surfaces.

While this technique is quite efficient, it isn't without its flaws. For starters, no matter how much care is taken or how soft a dusting brush is used, the print sometimes ends up getting damaged. Also, fingerprint powder doesn't work particularly well on porous surfaces, so forensic scientists use chemical means to turn the invisible visible.

Silver nitrate works wonders for developing latent fingerprints on cardboard and paper. Fingerprints are primarily made up of amino acids, lipids or oils, and sweat (water and salt). When silver nitrate is applied to the suspected object, it reacts with the salt (sodium chloride) left behind after evaporation of sweat and forms colourless silver chloride, which, when exposed to ultraviolet light, produces a reddish-brown or black colour. Thus, it reveals a well-defined friction impression that can be photographed for evidence.

Ninhydrin is another chemical that can be used to detect fingerprints on surfaces such as walls. When a 0.6 per cent solution of this compound is sprayed onto a surface, it can cause the area containing a fingerprint to turn purple. This colour change occurs because ninhydrin reacts with amino acids present in sweat, producing a compound known as Ruhemann's purple.

For rough, non-porous surfaces like textured plastic or metal that don't absorb liquids, superglue or cyanoacrylate fuming is preferred for fingerprint development. The suspected evidence is placed inside an enclosed chamber which is slowly filled with superglue fumes for a couple of hours. The cyanoacrylate ester molecules in the glue react with skin residue and polymerise to create a 3D matrix, which materialises as a white-coloured fingerprint.

I know we are curious to discover how these prints are analysed, but before we go there, here's one last fingerprinting technique that can produce high-quality fingerprints in less than fifteen minutes. Vacuum metal deposition (VMD), originally developed to coat glass for mirrors, is now a powerful technique for developing fingerprints, particularly on non-porous surfaces such as plastic bags. The process involves evaporating thin layers of gold and zinc onto a surface inside a vacuum chamber. The gold penetrates the oily residues, leaving clean areas for zinc to deposit. Since zinc won't stick to the oil-covered gold, it highlights the ridge patterns. Now onto the analysis.

How are fingerprints analysed?

On television, a forensic expert enters the lab with a fingerprint on a glass slide, places it inside a machine that scans an image of the fingerprint a couple of times, and after cycling through a range of misses a result flashes up – 100 per cent match! If only fingerprint analysis were this easy in real life.

It is a lengthy process that begins at the crime scene, where the lifted fingerprints are meticulously photographed for permanent record-keeping and to establish their location. The lifted prints or their photos are then sent to a forensic laboratory, where fingerprint experts take over for the next step, which is the examination process following the ACE-V (analysis, comparison, evaluation and verification) method.

Not every fingerprint collected from the crime scene is usable. The analysis step involves evaluating whether a fingerprint has sufficient quality and details necessary for the comparison and matching.

For comparison, fingerprint experts rely on Galton Points or minutiae – primarily ridge endings and bifurcations – to identify individuals. While features like lakes, islands, spurs and crossovers exist, they're often seen as variations or combinations of the two main types. Examiners also consider finer details, such as sweat pores and subtle ridge textures, to strengthen their analysis.

On locating the minutiae on a fingerprint, the experts then mark its location and orientation, and record its type. To compare prints, the experts print the fingerprint obtained from the crime scene on a transparent sheet and then overlay it onto a marked paper copy from a known individual. By aligning the minutiae, they visually check for matching features to confirm a possible identity.

This was the standard comparison process until the twenty-first century. Today, most countries use AFIS (automated fingerprint identification systems), advanced biometric systems that store millions of digitised fingerprint records. These systems are powered by algorithms designed to compare fingerprints by mimicking what human examiners do, looking for the identifiable Galton Points.

After the initial screening is carried out by AFIS, the possible matches undergo evaluation by a forensic examiner. They closely examine the similarities and differences between the prints and reach one of three possible conclusions: the prints are from the same source, they are from different sources, or there isn't enough detail to make a definitive determination.

The final step is verification, where a second independent examiner repeats the entire process to confirm or refute the first examiner's conclusion. Unfortunately, there's no satisfying loading bar or a dramatic '100% match' notification at the end. Disappointing, I know!

Evidence with a double helix twist

We probably don't need to begin this section with a statement explaining that DNA is the blueprint of life, which contains all our genetic information, or that its full form is deoxyribonucleic acid. As with the overused phrase 'mitochondria are the powerhouses of the cell', most of us are already familiar with what DNA is and what it does.

We are going to jump into the fact that almost every cell in the body contains DNA, and we constantly shed it wherever we go. We leave behind a traceable trail of our presence in any place we've been, shedding DNA through dead skin cells, hair, nails, fingerprints or bodily fluids, and so does a criminal. If it were 1983, the bad guy wouldn't break a sweat knowing this, but an accidental discovery by a British geneticist in 1984 has since ensured that every criminal is extra cautious about leaving behind DNA, because even the tiniest trace can lead straight back to them.

Alec Jeffreys joined the University of Leicester as a lecturer in the Department of Genetics, with an interest in uncovering how DNA varied between people, so he could trace how different versions of genes linked to traits and diseases were inherited through generations.

DNA is a large and complex molecule made up of three main parts: a phosphate group, a sugar group, and combinations of four nitrogen bases. These bases pair up in a specific way – adenine (A) always pairs with thymine (T), and guanine (G) always pairs with cytosine (C). The DNA double helix is formed when these building blocks link together in a chain, with the sugar and phosphate parts forming the backbone of the strands and the base pairs the links between the twin helices.

To understand inherited variations in DNA, Jeffreys began investigating minisatellites, which are short sections of DNA where sequences of base pairs (10–60 pairs) repeat over and over and over in specific regions of the genome.

For Jeffreys, 10 September 1984 began like any other day in the life of a researcher, picking up from where he'd left off the previous day. Jeffreys was looking at an X-ray film he had developed for his recent experiments: that's when he noticed that each of the human DNA samples he worked with had its own unique pattern of minisatellites, like a personal QR code.

He continued to examine more and more samples and found that it seemed true: every human has a unique DNA fingerprint, excepting twins. This discovery gave rise to one of the most robust and reliable tools in forensic investigations – DNA profiling.

How does DNA profiling work?

After the CSIs have identified the spots and sources for DNA collection, they follow strict procedures to preserve the quality of DNA evidence. Beyond carefully packaging weapons for DNA analysis, CSIs work methodically to collect every possible trace. They use sterile swabs to gather bodily fluids, cut out bloodstained pieces of fabric, scrape dried biological material from surfaces, lift trace evidence such as skin cells with adhesive tape, and pick up visible clues, like hair or nails, using clean forceps.

After the samples are carefully packed, labelled, sealed, photographed and documented, they are sent to the lab for DNA analysis.

Once in the lab, the process of profiling begins with extracting the DNA from the evidence samples. Most samples often do not have enough DNA to carry out multiple tests, so scientists use a polymerase chain reaction (PCR) machine that can create millions to billions of copies from a small initial sample.

Before we get into the analysis part, let's get a DNA fact straight. Unlike fingerprints, which are visibly unique in every individual, humans share 99.9 per cent of their DNA with each other. We share a majority of our coding DNA, which is responsible for building fundamental proteins essential for staying alive. Out of the 3 billion base pairs in each cell, the unique aspect is limited to approximately 3 million non-coding DNA bases, also known as junk DNA.

The scientists focus on a telltale region of non-coding DNA called the short tandem repeats, or STRs, which are short units of DNA, around four or five bases long, that are repeated numerous times in a row. These sections are often marked with special chemical markers before the PCR process. Once the copies are made, the markers are separated using capillary electrophoresis, which uses an electric current to sort the smaller and larger DNA pieces.

STRs are known to repeat multiple times at specific locations, known as loci. The number of repeats at a given locus varies from person to person, making STRs highly effective markers for human identification. Because humans inherit one set of chromosomes from each parent, every person carries two sets of STRs with a different number of repeats. This adds another layer of specificity to an individual's genetic profile, making STR analysis a robust DNA fingerprinting technique.

How are the samples matched?

Imagine the analyser has three different DNA samples: one and two are evidence collected at the crime scene, and three is a sample collected from a potential suspect. They choose to focus in on two loci or STR spots from the same DNA strand for each of the samples. Let's call these points A and B. Using an STR analyser, the lab scientist decides to calculate the number of times the sequence GATA repeats in each of the samples at locations A and B.

The first sample shows nine repeats of GATA at spot A and five repeats at spot B; the second has ten and eight, while the third again has nine and five. Since the number of repeats at a given locus varies from person to person, there is a very high chance that the first and third samples belong to the same person, suggesting they might have been present at the crime scene and could be a suspect.

What would happen if the DNA profile didn't match with any other sample collected by CSI? Then, analysts run the profile through DNA databases similar to AFIS. The UK has one of the longest-running forensic DNA databases in the world, known as the National DNA Database (NDNAD), which contains over 7 million individual person profiles and 600,000 crime scene profile records. The FBI in the USA also has a Combined DNA Index System (CODIS) with over 18 million offender profiles, 15 million arrestee profiles, and 1 million forensic profiles.

Forensic experts can also cross-reference the DNA profile from the crime scene with one of these databases to see if there is a match.

Albert DeSalvo, long suspected of being the Boston Strangler, was definitively linked to the 1964 murder of Mary Sullivan – long considered the last victim – through DNA analysis nearly 50 years after the crime. Initially, seminal fluid was recovered from the crime scene and preserved. As DNA technology advanced, investigators used Y-STR (Y-chromosome short tandem repeat) analysis to compare DNA from the crime scene with DNA from DeSalvo's living nephew, establishing a strong familial match through the paternal line.

To confirm this result, authorities exhumed DeSalvo's body in 2013 and extracted DNA directly from his teeth and femur. Nuclear DNA analysis conclusively matched DeSalvo's DNA to the biological evidence from the crime scene, leaving 'no doubt' he was responsible for the rape and murder of Mary Sullivan. This breakthrough relied upon advances in forensic DNA technology and was funded in part through the National Institute of Justice's cold case initiatives. The case remains a landmark example of how modern DNA testing can help solve crimes that have gone unsolved for decades.

Blood, bugs and botanicals

Your freshly vacuumed living room might look pristine to your naked eye, but if you let a forensic serologist loose, they can extract enough evidence to link it to you and every guest from your party last weekend.

Forensic serology is the field of forensics that utilises laboratory techniques to detect and analyse bodily fluids, including blood, semen, saliva, sweat and urine. This field involves detecting and often identifying body fluids at crime scenes, with tests ranging from blood typing and stain characterisation to blood spatter analysis for crime reconstruction, paternity testing, semen detection in sexual assault cases, and DNA profiling to pinpoint victims or suspects. In this section, we'll focus on just one fluid: blood.

DNA analysis of blood

Red and viscous, all human blood might look the same, but it is not. It comes in different types, which are determined by the presence or absence of certain antigens – substances that can trigger an immune response if the body identifies it as foreign. There are more than 40 different blood grouping systems, the most widely known being the ABO system.

In the early 1900s, Karl Landsteiner, a scientist at the University of Vienna, was studying blood transfusions and why some people have severe reactions after transfusions. He started experimenting and observed that sometimes the serum from one person's blood caused the red blood cells of another to clump together, while in other pairings this did not occur. Based on his observations he identified three distinct blood groups – A, B and C, which was later renamed O. A few years later one of his students identified the fourth type, AB.

Type A individuals have the A antigen.
Type B individuals have the B antigen.
Type AB individuals have both A and B antigens.
Type O individuals have neither A nor B antigens.

This discovery not only greatly enhanced the safety and effectiveness of blood transfusions, but also gave medico-legal experts a valuable new tool. Blood groups are considered genetic markers, as their inheritance strongly follows the Mendelian laws and remains unaffected by environmental factors, diet or disease. Forensic scientists now have a definite way to determine if the blood found at the scene belongs to the victim or the suspect, based on whether the pattern of clumping reactions matches that of the blood collected from the victim or not.

Blood spatter analysis

Blood is our identity; it carries the biological fingerprints of the families we are born into. It nourishes all our cells with the fuel required to keep the body up and running, but when someone wants to harm or kill a human, all they have to do is provide a way for the blood to escape the skin-wrapped vessel and let it flow.

Violent crimes involving assault, stabbing and gunshots leave behind a trail of spatters, smears and drag marks, among many other unique marks left behind by interactions between physical forces and blood.

Blood is a consistent fluid made up of liquid (55 per cent) and solid (45 per cent) parts. It is non-Newtonian in nature, meaning its viscosity reduces and it becomes significantly thinner and travels faster than usual upon encountering impact. Blood always finds a way to create a pattern that carries with it clues about the point or manner of exit from the body. Analysts rely on these clues and crime scene reconstructions to trace the forces behind the stains and piece together the sequence of events that unfolded during the crime.

One bloody axe, two beds, three bodies and a survivor who is an unreliable eyewitness, or a killer. The police were called to the scene of a triple homicide: a young mother and her two children, aged six months and two years, were murdered with an axe. The only surviving member of the family was the husband, who said he saw what his wife had done to the children, so he killed her with the same axe she had used.

To establish the sequence of events, the experts looked at the evidence: three victims lying on separate beds in different rooms, blood pooled on the beds and tiny splashes across the ceilings, walls and even on the victims. If the wife killed the children with the axe, she would have splashes of their blood on her, so the experts carefully collected blood spatters from the husband and the three victims.

DNA analysis revealed that the only blood found on the woman was her own – none from the children – which ruled out her killing the children with an axe. On the other hand, the blood spatter stains collected from the husband's clothing were from all three victims. Upon being interrogated with the evidence in hand, the husband revealed that he first killed his wife and then his children. Establishing a proper sequence of events is essential, as it can sometimes influence the sentencing by the judiciary.

The origin of this field of study is as gory as the crime scenes that require blood spatter analysis. It began with bashing rabbit heads with a hammer. In 1885, Eduard Piotrowski, an assistant at the Institute for Forensic Medicine in Poland, wrote a paper titled 'Concerning the Origin, Shape, Direction and Distribution of the Bloodstains Following Head Wounds Caused by Blows'. To gather data for this study, he placed live rabbits in front of large paper sheets and bashed them with a hammer, rocks and hatchets from different angles. He hired a painter to make accurate illustrations of the blood patterns left behind.

Times have changed, and scientists simulate blood spatters using defibrinated sheep or pig blood, where the plasma fibres responsible for clotting are removed to ensure that the blood remains fluid during testing. But one of Piotrowski's findings that still holds true is that in most cases, bloodstains appear with the second blow. In other words, when struck with a sharp or blunt object, the first blow creates an open wound, and the blood sticks to the weapon. It's only when the assailant withdraws the weapon and goes for the second blow that the motion sets in a series of blood spattering events. First, the blood that is stuck to the weapon flies off the surface, creating patterns. When the weapon lands on the open wound for the second time, the blood displaced by it is then cast off to the surrounding surfaces.

The bloodstains aren't random; each droplet carries a clue in the form of its shape, size, position and direction. The blood patterns found in a crime scene are often categorised into three basic types.

The first broad category is low-velocity blood patterns, which are formed by a minimal amount of force or even the passive influence of gravity. For example, blood dripping from the tip of a knife or blood oozing out of the victim's body as they move to call for help. The blood droplets formed in these cases are larger and result in stains greater than 4 mm in diameter. This category also includes blood-into-blood scenarios where blood dribbles into an existing pool of blood, giving rise to characteristic satellite patterns. These patterns occur when blood bounces off a large pool of blood, creating tiny splashes, and can reveal the positions of the victim and assailant, as blood falling from different heights forms distinct satellite patterns. For instance, stains from a standing victim's stabbing differ noticeably from those where the victim is seated.

As the velocity increases, blood travels further through the air, breaking into finer droplets that leave smaller stains. The second category, medium-velocity impact spatter, occurs when a blood source is struck with the force of a blow from a baseball bat, knife or hammer. The stain diameter ranges from 1 to 4 mm. Medium-velocity patterns can distinguish a self-defence attack from an intentional one. Drawing from Piotrowski's observations, it requires more than a single blow to produce the spatters under this category. A person being robbed might bash the burglar's head once to incapacitate them enough for them to escape to safety, and the first wound-causing blow barely causes any spatter. However, hitting someone's head repeatedly to create enough spatter on the nearby wall or clothing does point to intentionality.

Now, coming to the high-velocity impact, blood spatters that are caused by forces so strong, such as a gunshot, that they almost aerosolise the blood into fine droplets that are often hard to register with the naked eye.

Analysing the spatters found at a scene where a firearm was discharged involves clear blood trails that can help determine where the shooter and the victim were when the incident occurred. A gunshot is followed by a forward spatter, which is caused by the exit wound where the blood is travelling in the same direction as the bullet. This leaves smaller droplets spread over a wide area behind the victim. Then there's backspatter, also called impact spatter, where blood travels back in the direction the bullet came from. These spatters often form larger, heavier droplets, clustering densely in the areas right next to the point of impact.

These blood patterns help piece together the puzzle of what exactly happened at the crime scene, but what if the tell-tale signs turn into red herrings?

Rod Englert, a blood spatter expert and crime scene reconstructionist, provides rather interesting examples of how certain clues can completely throw detectives off the track in his book *Blood Secrets*.

Flies on the wall of a crime scene with bloodshed can create an illusion of high-velocity spatter. The winged creatures love to lay their eggs in blood, so they walk around in the wet liquid, poop and puke in the area, and leave behind tiny dark specks that to the naked eye might look like mist from a gunshot. This is often taken to mean the killer must have been standing here. A quick look under the microscope can set things right by bringing into focus the mistaken blood spatters; they are indeed Q-shaped puke marks left behind by the flies.

Before we move on to reconstructing a crime scene, here are a few more blood spatter clues for your next scientifically accurate description of a crime scene for your detective novel:

1. If blood drips on a surface at a 90-degree angle, it forms a perfect circle, but as the angle becomes smaller, the stain left behind by the blood drop becomes elongated.
2. If you see crater-like gaps in the blood spatter, the blood source was an airway, like lungs or throat. Air mixed with blood creates bubbles, which pop over time and leave ring-shaped voids in bloodstains.

3. If the bloodstain is brighter red than the surrounding ones and has a spurting pattern, the blood source is probably the heart. The pumping nature of the heart and the presence of highly oxygenated blood give the stains their unique markers.

Sometimes, some cases push detectives to look beyond guides and textbooks and transport themselves into the crime scene by recreating it.

Crime scene reconstruction

Crime scene reconstruction is the science of piecing together the story a crime scene has to tell. To move forward in a case, reconstructionists often have to work backwards, from after the incident to during and before. It's the step-by-step process that answers the two main questions, 'what happened?' and 'how did it happen?', using crime scene observations and laboratory examination of evidence. It is not to be confused with crime scene recreation, often shown on TV shows, where the detectives place themselves at the crime scene and re-enact different aspects of the crime scene to duplicate the possible chain of events.

Let us assume investigators are responding to a homicide scene in an apartment and the evidence in hand is: a) a broken window with bloodstains, b) bullet casings and blood spatter inside, and c) witnesses who saw someone fleeing from the scene.

Reconstruction begins by collecting all the physical evidence, like gunshot residue, blood spatter, witness statements, documents, laboratory test results, and photographs of the evidence and layout of the crime scene, as well as the victim, in the case of a homicide. In our case it would be documenting the bloodstains, collecting the bullet casings and glass shards, and photographing the victim and the crime scene.

This is followed by conjecture, in which reconstructions typically involve walking through the crime scene and attempting to think like the offender to develop a realistic explanation of the events surrounding the criminal act. It is not the only explanation for why things happened, but rather their best estimate among many possible scenarios. Was the window broken before or after the shooting? Did the assailant fire from inside or outside?

Then comes the time to formulate a hypothesis, which in this case is an informed guess based on additional information gathered through laboratory analyses of physical evidence. The results from DNA and fingerprint analysis, bloodstain interpretation, examination of gunshot residue, etc. are brought together to form a hypothesis of the probable chain of events.

Analysing the glass fragments and blood patterns, in our case, indicated that the victim was wounded during window breakage, and the ballistics report suggested that the shots were fired from inside.

The penultimate stage is testing, where the formulated hypothesis is rigorously scrutinised to challenge or confirm its validity. This may involve matching crime scene evidence to known reference samples, carrying out chemical and microscopic examinations, all aimed at either reinforcing or disproving the interpretation. In our homicide case the blood on the glass matched the victim, the firing position was approximated by trajectory analysis and the witness accounts aligned with the existing ideas.

At this stage, if forensic scientists find that the test results do not align with the hypothesis, then it's back to the drawing board.

After all the evidence is analysed and validated, reconstructionists piece together everything they know about the case to formulate the most plausible theory. This theory is then documented alongside supporting evidence and presented to the courts.

A theory based on all the evidence and testing might state that the assailant broke the window, entered the apartment, and fired upon the victim. The physical evidence – the direction of shattered glass, blood patterns and ballistic analysis – corroborate this sequence.

Effective crime scene reconstruction can not only catch a killer but also help the innocent regain their freedom.

The latest 3D crime scene reconstruction technology uses laser scanners and computer modelling to create a virtual replica of the scene. Investigators digitally capture every detail, from bloodstains to bullet trajectories to body positions, and build 3D models similar to walkthrough video games to analyse the events as they happened. This immersive setup not only helps experts visualise complex scenarios but also comes in handy as show-and-tell evidence in court.

In 1994, 22-year-old Kei Sunada was shot to death in a fourth-floor stairwell of a Queens, New York, apartment building. Two men, Armond McCloud and Reginald Cameron, were convicted based solely on (coerced) confessions extracted after long, intense interrogations without legal counsel.

A forensic team from New Jersey Institute of Technology conducted a detailed crime scene reconstruction nearly 30 years later. They laser scanned the scene to map the building's fourth-floor stairwell and hallway digitally, integrated the data with original photos and case files, and performed bullet trajectory analysis.

Their reconstruction showed the shooting actually occurred in the stairwell, not near the elevators as police had claimed. The bullet trajectory placed Sunada inside the stairwell, standing near a door, with the shooter facing him. Ricochet marks and blood evidence supported the theory but contradicted the confessions. This reconstruction and re-evaluation of the witness statements exposed the false narrative previously created and helped clear the names of two people who spent over three decades locked up for a crime they didn't commit.

Investigators hold onto every single clue, from the most obvious ones to the smallest traces found at the crime scene, including the arrival of pesky flies and maggots.

A fly on the corpse is worth more than two on the wall

Once upon a time, a man was found lying on the roadside. A closer look revealed that he had been stabbed to death. The town judge/physician/coroner began his investigations. His first step was identifying the type of blade used to kill the person. After trying out several sharp objects on a cow carcass, he concluded that a sickle caused the stab wounds on the victim. He then interrogated the victim's wife and neighbours to get some leads on possible suspects and the motive. The victim supposedly had no enemies but was in an argument with a moneylender, as he wasn't able to repay the debt to the lender's satisfaction. The moneylender completely denied having anything to do with the murder.

The judge, however, knew some tiny informants who could lead him to the culprit. The next day, he called upon all the adults in the neighbourhood to stand in a line and place their sickles at their feet. Now, all he had to do was wait. As expected, he spotted a few winged visitors on the sickle that belonged to the moneylender. The coroner immediately knew he was the killer because flies are attracted to blood, even the tiniest amount, often invisible to the naked eye. The moneylender confessed to the stabbing, and forensic entomology successfully ensured that a criminal was caught.

The only catch is that this incident occurred long before anyone had even heard the words 'forensic' or 'entomology'. This use of insects to investigate a crime appeared in the thirteenth-century Chinese physician, judge and coroner Sung T'zu's book *The Washing Away of Wrongs*.

Even today, forensic experts turn to forensic entomology – the study of insects in relation to criminal investigations – as a vital tool for estimating the time of death, particularly when the circumstances of the crime are unclear or other evidence is lacking.

The phrase *dead body* may evoke a sombre image for us, but if insects could read, some of them would be more than overjoyed because, to them, decay is a goldmine of nourishment. The odour emanating from a corpse brings a variety of insects to the yard, depending on which stage of decomposition it is at. The first to visit the scene are blowflies and flesh flies. The moist microbe-rich environment of carrion is perfect for the female blowflies to not only feast on, but also for laying their eggs. In more mature stages of decay, the beetles arrive to feast on the body tissues and the maggots that have emerged from the blowfly eggs. Even certain species of wasps join the scene, targeting the maggots. The presence of these insects can enable forensic entomologists to assess the time of death.

With just blowflies and their progeny alone, investigators can piece together the timeline of death.

Minutes to 24 hours: Blowflies often reach a body within minutes to a day, laying eggs in wounds or natural openings.

1–10 days: This is the larval development stage where the eggs hatch into maggots and go through first, second and third stages of growth over several days. Each stage can be identified by the maggot's size and the number of breathing holes.

Days–weeks: In the pupation stage, the fully grown larvae leave the body to pupate in a dark, safe spot, turning into hard brown cases where they change into adult flies.

Weeks: Adult flies break free from the pupal case after several days or weeks, leaving empty shells that mark the end of a life cycle.

Temperature on and around the decaying matter plays a key role in dictating the rate of development of insects. A warmer day accelerates their development and a colder one slows it down. The temperature of the decaying site and the oldest stage of insect present can help calculate the day or range of days when the first insects laid their eggs, which directly links to the minimum time since death. In other words, if the oldest insects are ten days old, the person has been dead for at least ten days because these insects only land on dead and decaying matter. Once the first adults emerge, experts switch to tracking waves of new insect species to estimate deaths from about a month to a year old.

Blowflies are generally considered diurnal insects, which prefer to lay eggs during daylight hours when sunlight and warmth are present. However, some research suggests that under certain circumstances, such as warm nocturnal temperatures or the presence of artificial light, blowflies can sometimes lay eggs. Since more experimental evidence is required to back this view, forensic scientists, for now, take into account the fact that the files prefer to lay eggs during the daytime, a behaviour which might shift the timeline of death by approximately twelve hours.

Insects also act as witnesses to a poisoning or overdose. When they feed on a decomposing body, drugs or poisons present in the tissues enter the insects' bodies and become trapped in the insects' outer shells or cocoons, much like how hair traps chemicals in living humans. Since these structures are made of chitin, a tough and chemically stable material, it can preserve traces of drugs, like cocaine or opiates, and poisons, such as mercury, for years after death – even when human tissues have completely decayed.

It is not uncommon to come across people finding insects in their food and seeking compensation from the restaurant. Social media and Google reviews aren't kind to places where such incidents occur, and rightly so. This has given rise to a BYOI (bring your own insect) scam, where people plant insects in food or its surroundings to blackmail business owners. But not to worry – forensic entomologists have got their backs.

A hotel proprietor in Yorkshire, United Kingdom, was approached by the father of a bride whose wedding reception had been hosted at his premises a few days prior. The father had pictures from the wedding day where one could spot what looked like cockroaches on the tables and walls. He asked the owner for compensation of £3,000 or he would show the pictures to health inspectors. The owner, sure of his hygiene practice, refused to pay the blackmail and called the police.

The police sent the photographs to Dr Zakaria Erzinclioglu, a British forensic entomologist, for a closer examination. The first thing that struck him as odd was that cockroaches usually don't appear in crowded, well-lit rooms; they prefer dark, quiet spaces. Upon enlarging the image, he realised they weren't cockroaches at all, but mites, and judging by the scale of surrounding objects in the photo, their size would be measured in centimetres. Mites, however, are barely one millimetre in length. Dr Erzinclioglu's analysis confirmed that the images were forged, and this was clearly a case of blackmail.

Experts also study the subtle traces insects leave behind – such as the marks from their regurgitation left around blood spatters. Scientists have even been able to extract DNA from blood that has passed through a fly's digestive system, sometimes revealing the identity of a killer or a victim. Another example is the mosquito we met in the introduction, which led the detectives to the car thief.

It is often said that nature is a gift that keeps on giving, which stands true for forensic clues too.

From the leaf sticking in the victim or criminal's clothing to the wooden murder weapon to the nettle growing near a clandestine grave, these clues have provided forensic botanists with enough information to seal the fate of a criminal. Apart from murders, forensic botany can also prevent illegal trading of wood and plant species.

In 1996, two tourists strolling through a park in the town of Noosa Heads, Australia, discovered the brutally murdered body of Samantha Bodsworth, a mother of two. The primary suspect was her former partner, Michael Bodsworth. When interrogated about his whereabouts at the time of the murder, he claimed he hadn't seen her that day, and the police couldn't do anything, as they had nothing tying him to the murder scene. However, Michael's days as a free man were limited, thanks to pollen, a powdery substance found in flowering plants that is biologically designed to be sticky, which placed him around the crime scene.

The area where her body was found had numerous flowering wattle trees, a species native to Australia. Her car, which was found outside a nightclub near the suspect's house in Gympie, was also littered with wattle flower petals and plant debris in the driver's and passenger's seats.

The challenge for investigators was that the crime scene contained two species of wattle, and wattle trees were also growing near the suspect's home. So they turned to Dr Lynne Milne, a forensic palynologist, hoping she could help untangle the case. Lynne noticed that the wattle flowers growing in the park and near Michael's home were of a different species, and pollen is unique to every plant species.

Getting up close with the pesky pollens via light and scanning electron microscopy, Lynne was able to successfully identify that the pollen on the suspect's clothing matched the pollen from the crime scene and not from the trees which grew near his house. Lynne's analysis and the police investigations got Michael arrested, convicted of the murder, and sentenced to life imprisonment.

Forensic botany, the application of plant science in criminal investigations, can also provide essential clues for locating clandestine and shallow graves. Across much of South America, thousands of people have vanished, many of them victims of forced disappearances. In Colombia alone, about 68,000 people are missing, with more than 20,000 believed to have been taken against their will. The graves of these victims can be found hidden in all kinds of environments. Researchers from Colombia carried out experiments simulating clandestine graves, observing changes in plant growth over graves, which can serve as visual clues to the presence of buried remains.

In Western European countries, nettle and early-flowering perennials grew in areas where the primary vegetation was disturbed or replaced for burials. Similarly, the Colombian study found that the type of plants growing around the grave varied depending on the age of the carcass. After four weeks, the ground between the simulated graves was mostly covered in Kikuyu grass, while the graves themselves had little vegetation. By eight weeks, wild radish began growing over the graves. After sixteen weeks, all graves were covered in lush wild radish, while Kikuyu grass still dominated the areas in between.

Stopping illegal timber trafficking

Every year, acres of forest are cleared to fulfil the demands of the illegal timber trade. Rosewood, a richly hued hardwood with beautiful veining, is in high demand in the luxury furniture industry, the only problem being that it is an endangered species that grows in the Amazon region and certain parts of Africa and India. The Convention on International Trade in Endangered Species (CITES) has placed severe trade restrictions on rosewood to prevent the plant from going extinct. However, high demand in international markets has fuelled an underground network of wood smuggling. Between August 2022 and July 2023 alone, about 311,000 acres of Amazon woodland was illegally cleared. A similar story is unfolding in eastern Russia, where oak logging is banned. Yet forests continue to be targeted, threatening the last remaining habitats of wild Siberian tigers and Amur leopards.

Customs checks, dodgy import/export documents or even tip-offs can raise questions regarding the legality of the timber being imported. When in doubt forensic experts can extend a helping hand.

Just as every human carries unique DNA, each plant species has its own genetic signature. To tap into this, scientists use a technique called DNA barcoding, which relies on a short, standardised DNA sequence from the chloroplast (the part of the plant that contains chlorophyll) to accurately identify the species.

Scientists also exploit the small molecular differences in plants that reflect their genetic diversity and their place of origin for species identification. They use a tool called DART-MS (direct analysis in real time mass spectrometry), which ionises molecules from wood samples and produces a spectrum showing the most abundant ions and the chemical fingerprint in each sample. By comparing these spectra with a reference database, investigators can accurately identify the wood's species.

Forests are home to millions of creatures big and small, and crime against nature and its children needs as serious action as crimes against humans. Forensic botanists, aka nature detectives, can help authorities worldwide track down illegal deforestation and hold accountable those who profit from destroying our planet.

Ballistics and arson

Nothing is perfect, and this applies not only to humans but also to objects. Even the smoothest surfaces and the most perfectly moulded pieces of chocolate have flaws that make them unique when observed at a microscopic level.

When a gun is manufactured, its metal components are shaped, ground and finished by machines. Despite being manufactured in the same facility, not every gun is perfectly identical, because no two machining tools are ever exactly alike, nor are the metal parts that are being worked on. All these slightest of differences leave behind microscopic ridges, scratches and grooves on the metal surfaces of the weapon, which are known as tool marks.

A bullet cartridge is made of four parts: the casing; the primer, which serves as the ignition system; the propellant, traditionally gunpowder but these days a nitrocellulose-based smokeless powder; and the projectile, commonly called the bullet. When the firing pin of a gun strikes the primer, it sparks the propellant, producing hot gases that expand rapidly inside the case. This surge of pressure pushes the projectile into the barrel which then bursts out and shoots towards the target.

Due to the immense pressure created inside the barrel of a firearm when discharged, the tool marks often get imprinted onto the bullets and the casings. The barrels of many rifles and handguns have spiral grooves called rifling, which puts the bullet in a spin before it leaves the gun. This spin provides the bullet with stability and enhances its accuracy. These rifling patterns, which are different for different guns, get imprinted on the bullet during firing, creating unique indents, which along with other tool mark form what is known as bullet fingerprints.

Crime scene investigators collect the spent bullet casings, bullets and, if lucky, the gun, and send them to forensic firearm experts. Armed with their trusty microscope – and, more recently, 3D scanners and high-resolution imaging gear – experts can identify these minute imperfections and match the bullet with the firearm it was released from.

Mob wars weren't uncommon during the Prohibition era in the USA. The alcohol ban, which was driven by the desire to reduce the social evils associated with alcoholism, gave rise to a black market where competition was fierce and violence was deemed the only logical way to deal with disagreements. In Chicago's underworld, Al Capone, aka Scarface, was locked in a relentless and bitter rivalry with George 'Bugs' Moran that led to a cold-blooded massacre which left the US completely shaken.

On 14 February 1929, when many Americans were celebrating Valentine's Day with their partners, a car similar to the type used by police squads in Chicago pulled up in front of a garage. Four men, two wearing police uniforms, got out of the car and were seen carrying what looked like riot guns (to the neighbours). The men entered the garage where seven members of Moran's team were gathered, and asked them to line up against the wall. A few moments later, the area was filled with the deafening noise of gunshots. The St Valentine's Day Massacre left behind a room full of bloodshed, bodies and bullets.

Calvin Hooker Goddard, a former US Army Medical Corps officer and ballistics expert, who later became known as the 'father of forensic ballistics', was invited to Chicago and offered a larger laboratory than the one he had in New York. In return, he was expected to help the Chicago police find the culprits. All their investigative leads had gone stagnant, but there was hope that the spent bullets and casings could lead the way.

Upon his arrival, Goddard was handed a huge collection of bullets and shell casings, including 70 .45-calibre cartridge shells that were carefully collected from the crime scene by Cook County Coroner Herman Bundesen.

His first goal was to identify the weapon used. To compare tool marks on reference bullets and cartridges with those on evidence exhibits – to determine whether they were fired from the same firearm – he used a comparison microscope. This specialised microscope consists of two microscopes connected by an optical bridge, allowing a split view that makes it possible to examine two samples side by side. He conducted test firings with several firearms, studied the markings and impressions left on the bullets and casings, and compared them with those recovered from the massacre.

He concluded that the .45-calibre shells had been fired from Thompson (or Tommy) guns. He checked the Thompson guns owned by the Chicago police department and found that none of them had fired the bullets from the massacre.

A few days later, two Tommy guns were seized from a different violent crime investigation that reached his lab.

The tool marks on these weapons matched those recovered from the crime scene. Investigation revealed that Fred 'Killer' Burke, the man involved in that case, was among the suspects in the massacre, helping solve the case.

Firearm and bullet fingerprint analysis is a part of internal ballistics that focuses on what happens inside the weapon when it is fired, but that's not always enough to demystify a crime scene. What happens when the bullet leaves the firearm and shoots through the air towards the target, and what it leaves behind when it hits the target, also provide essential insights.

External ballistics focuses on the bullet's journey after it leaves the barrel and is in flight, or on its trajectory. This trajectory is rarely a simple straight line; it's shaped by external factors like gravity, spin, air resistance, velocity and even wind direction. By analysing the bullet's trajectory, forensic experts can determine the origin point, angle and firing range of the gun used, all the details essential for reconstructing the crime scene, and identifying the shooter's location.

Trajectory analysis begins with finding the bullet hole in the crime scene, since the shape of the hole created can tell us about the angle at which the gun was shot. If the bullet hole is perfectly round, which is quite rare, the gun was shot perpendicular to the target. Even a slight shift from 90 degrees leaves behind an elliptical hole. The angle at which the bullet hits the target is called the impact angle. To determine this value, analysts first measure the major axis (typically the vertical line) and the minor axis (the horizontal line) of the elliptical bullet hole. These values, when placed in trigonometric functions, give us the impact angle.

Forensic ballistics examiners work in a similar fashion to this day. What seems like an ordinary piece of brass on the ground can therefore become one of the most powerful forms of physical evidence in a criminal investigation, narrowing down suspects and, in many cases, pointing directly to the weapon that was used.

To map a bullet's flight path, investigators often use simple stringing or more high-tech laser methods. In stringing, rods are placed in bullet holes, and lengths of string are stretched between them, creating a clear visual representation of the trajectory. Lasers can trace a cleaner, straighter line, although they sometimes require smoke or low light to be visible indoors.

In more complex cases, investigators turn to tools like photogrammetry, which can capture spatial data of the crime scene. These methods allow them to recreate bullet trajectories in detailed 3D, producing powerful visualisations and simulations that bring shooting scenarios to life.

Forensic experts also take the help of ballistic tables, which are reference charts that predict how a bullet will travel once it leaves the muzzle. These references are created over years and years of data collection from multiple cases, and take into account gravity and the other important factors that can influence a bullet's path.

Once the bullet reaches its final destination, terminal ballistics comes into play. It focuses on what happens when a projectile hits its target and transfers its energy. Understanding how bullets damage living tissue and other surfaces upon impact can help explain where the shooter might be standing or the direction from which the bullet was fired.

For instance, when a bullet enters the body, it usually leaves behind a small, neat wound with a ring-shaped scar where the projectile scraped the skin. If the shot was fired at close range, forensic analysts might also see soot, stippling, or gunshot residue around the point where the bullet first made contact with the body.

On a sidenote, gunshot analysis also includes testing for microscopic particles released when a gun is fired. Tiny bits of bullet and components of its cartridge can find their way onto hands, clothing or nearby surfaces. By testing for gunshot residue (GSR), investigators can tell if someone fired a weapon, handled it, or was simply close to the gun when it went off. For example, if a suspect claims they were nowhere near the crime scene but gunshot residue is found on their sleeves, there is a very high chance that they are lying.

Exit points are far messier than entries. On hard surfaces like glass or furniture, they push material outward, leaving behind splintered wood or radiating cracks in glass. The same principle applies to the human body: as the bullet travels through, it deforms under internal resistance, tearing more tissue on its way out. Unlike entry wounds, exit wounds carry no soot or powder burns.

By combining the information gathered from wound features and gunshot residue tests, analysts reconstruct the bullet's journey and pinpoint where the shooter was standing.

Forensic ballistics, however, faces some criticism due to its weak scientific foundation. The core idea behind tool mark analysis is that every gun leaves a unique, fingerprint-like impression on the bullets it fires. However, many critics, including the National Academy of Sciences (NAS) of the USA, state that this fundamental assumption has not been fully proven through extensive peer-reviewed testing. More research is needed to determine the extent to which these results are truly unique and to assess whether the methods can consistently deliver reliable and repeatable answers.

Wrongfully arrested and saved by ballistics

In 1993, Patrick Pursley was convicted of murder solely based on forensic ballistics evidence, since no other form of evidence was available that could link him to the crime scene. At trial, the prosecution's ballistic expert testified with strong certainty that tool marks imprinted on the bullets from the crime matched those from Pursley's gun, seized from his home.

In the 23 years he spent in jail, Pursley consistently maintained his innocence and wrote to many organisations to help him overturn his wrongful conviction. In 2001, Pursley discovered the Integrated Ballistic Identification System (IBIS), which utilises advanced high-resolution, multidimensional imaging to analyse and compare fired bullets and shell casings with greater accuracy than the methods available in the early 1990s.

Pursley pushed for a law amendment that would allow re-testing of ballistic evidence from his case. After the law change in 2008, Pursley's new legal team, led by Northwestern's Center on Wrongful Convictions, filed the first motion of its kind in Illinois for ballistic comparisons with the National Integrated Ballistics Information Network (NIBIN).

After several years of requests and denials, their petition was heard, and the Illinois State Police finally uploaded the ballistic evidence into NIBIN. New ballistic testing found no conclusive match between the crime scene bullets and Pursley's gun.

Just before the retrial, new information came to light. The investigators never actually recovered a gun at the crime scene, so they simply inserted Pursley's own gun into the evidence record.

In 2019, after 23 years behind bars, Pursley was finally exonerated, his innocence confirmed with the help of a forensic technique that had initially failed him.

Digital fingerprint

Charlotte Robinette was packing her lunch in her home when the phone rang.

'Mom, please help me. They've taken me,' her daughter's terrified voice pleaded. Then she heard a man's voice say, 'We've got your daughter. How much is her life worth?'

The kidnappers ordered Robinette to withdraw money from the bank and wire transfer money across Phoenix, USA, and Mexico, each demand harsher than the last. Sleep-deprived and panicked, she obeyed until they told her to cross into Mexico and deposit the last of her savings.

Exhausted and having paid up over $11,000, she just wanted to know that her daughter was alright. So she picked up her phone and rang her. To her surprise, her daughter picked up her phone and sounded perfectly alright, and she was at home. The horrifying truth hit her – she had been scammed.

Robinette had fallen prey to a virtual kidnapping. Criminals use spoofed numbers and personal details available on social media to extort money out of loved ones. Crime no longer hides in dark alleys or within bullet-ridden walls; the digital space has become an equally daunting battleground, the rise of generative AI only fueling it further.

Wherever there is crime, there must be forensics, and the digital world is no exception. In this section, we'll explore some of the most common procedures in digital and cyber forensics.

To begin any kind of investigation, forensics experts need to start with evidence. Digital evidence includes any data stored or transmitted through a computer that helps support or refute a theory of how a crime (online or offline) unfolded. The data could be in the form of emails, text messages, images, audio or video. Digital forensic experts work with digital systems on three different levels.

Firstly, the open computer systems that most of us are familiar with include laptops, desktops and servers, which typically feature hard drives, keyboards and monitors. A single file might not only contain incriminating content but also carry hidden properties, such as metadata – data that describes other data – that can tell investigators when it was created, by whom, or even whether it originated on another device.

Then we have communication systems, such as phone and internet networks, ranging from traditional landlines and cellular towers to email servers and cloud platforms. All of these act as silent witnesses to billions of communications every day. With the right tools and forensic expertise, these witnesses can be made to point towards the criminal mastermind.

Text messages, chat app text conversations and emails all leave behind digital footprints that help investigators reconstruct a message's journey by examining logs from intermediate servers or routers, revealing crucial details like who sent it, when and what was said. They can also recover information even from encrypted apps by extracting metadata like timestamps, read status and participant details.

Everything is smart these days, from watches to refrigerators to bulbs. What makes them smart is an embedded computer that can help you plan your shopping list for the week based on what's left in the fridge, but it also serves as a source of digital evidence. Navigation systems in vehicles can track a car's location, bank cards can record transaction logs and smart doorbells can capture footage of movements in the front yard. In fact, voice assistant recordings and doorbell footage have been used in homicide and burglary cases to place suspects at the scene of the crime.

Nothing is ever deleted

Between 1974 and 1991, a serial killer, Dennis Rader, who called himself the BTK Killer – short for his modus operandi of Bind, Torture, Kill – ended the lives of ten people in Kansas, USA. He would select victims at random, stalking them for weeks, and then strike. His method followed a sadistic pattern where he would bind his victims with ropes or cords, torture them psychologically and physically, and finally kill them, usually by strangulation.

One of his earliest crimes was the slaughter of the Otero family in 1974. Rader broke into their home on a January morning, held them at gunpoint and bound each family member. He strangled the parents, Joseph and Julie Otero, before turning to their children, eleven-year-old Joseph Jr. and nine-year-old Josephine.

For decades, this supposedly perfect father, husband, churchgoer and Boy Scout leader taunted police and the media with letters and poems, basking in the fear he created among the masses. After 1991, he went completely off the radar. No murders, no hubristic taunts. However, this didn't last long.

He fell back into his old habits of writing to the media and police after a Kansas newspaper mentioned that the public had forgotten the BTK killer. This time, things were different; the world had entered a digital era, and forensic techniques like DNA fingerprinting were much more sensitive and accurate. The police jumped back into action and collected the DNA from under a victim's fingernails, and started testing hundreds of local men for a possible match. Luckily they found help from the killer himself.

Among the things he sent out was a purple 1.44-megabyte floppy disk, to a local news channel, convinced that the information he included was untraceable. But it wasn't the files he left behind that sealed his fate; it was what he thought he had erased. Forensic investigators discovered metadata hidden in a deleted Microsoft Word document, revealing the words Christ Lutheran Church and showing that the last edits had been made by someone named 'Dennis'.

When we select a file and click delete, we expect it to vanish from the face of the earth, but fortunately – or unfortunately – that is not what happens. The file system typically marks the space the file once occupied as available and erases the references required to locate it within the system. The actual data simply sits in the storage device, untouched, until we overwrite it with new data.

To recover such deleted content, the experts first make a bit-by-bit copy of the digital storage device to create a forensic image. Unlike a simple file copy or transfer, a forensic image captures everything on the device, including deleted files, hidden data and even the empty space that can be used to trace past activities. The goal is to preserve the original evidence in its exact state and yet have something for the investigators to analyse without any damage to the original.

Forensic investigators use specialised recovery tools to hunt for deleted files within a forensic image. These tools look for recognisable file signatures, called magic numbers, which are used to identify or verify the content of a file and are usually inserted at the beginning of the file. They also scan unallocated space, which may still hold fragments of deleted data, and examine file slack, which is leftover bits of information stored in the unused portions of clusters. In addition, some file systems keep journals or metadata logs that record traces of activity, offering valuable clues about files that were deleted but not fully erased.

After the forensic experts were able to find a name, the police, through a court order for medical records, were able to collect the DNA of the primary suspect's daughter, who'd just had a Pap smear at a Kansas hospital. There was an undeniable match.

After evading the law for 30 years, Rader was arrested in 2005 and put behind bars because of his vanity and a floppy disk.

Is it really hidden?

Steganography is the art of hiding secret data within another innocent-looking file, like embedding a message inside an image, audio or video file so the presence of the message isn't obvious. Unlike cryptography, which locks content in a visible but unreadable form, steganography conceals the fact that a secret exists at all. Its primary function is to covertly communicate information that a third party cannot intercept.

The act of steganography itself is legal, as it has genuine uses, such as watermarking for copyright protection, embedding patient information in medical images, or securing sensitive communications in defence settings.

In the wrong hands, however, it is a powerful tool for carrying out illicit activities while hidden in plain sight. Cybercriminals hide malware, also known as stegomalware, facilitate data exfiltration, or conceal illegal content like child pornography.

In 2010, the FBI apprehended a group of people who were embedding secret messages inside images posted on public websites. According to the US Department of Justice's charges, the alleged Russian spy ring passed on secret instructions to their handler in Moscow using steganography.

A common method for hiding a message in an image is called Least Significant Bit (LSB) steganography. It works by altering the tiniest details of a media file that are too small for the human eye to detect but are evident enough for a person on the receiving end of the message. For example, an 8-bit image uses 8 bits to define each pixel. If a pixel is

10110110, changing the last two bits from 10 to 11 makes it 10110111, a change that is invisible to the eye but sufficient for storing hidden data.

To an untrained eye, an image with steganography looks no different from the selfie you just posted on Instagram. But to an expert, subtle irregularities can give it away. Hidden data often leaves behind unusual patterns or distortions, whether in images, unused disk space or file partitions. Analysts start with simple checks using a common method, known as the known-cover attack, which compares an original image with its altered version to identify differences. They then move on to statistical tests that identify abnormal bit patterns.

In images, repetitive patterns may expose the fingerprints of a steganography tool. Over time, such comparisons accumulate signatures that not only confirm the presence of hidden messages, but also pinpoint the specific tool used to embed them, even when the original cover image is unavailable.

Forensic experts have at their disposal high-tech steganalysis software, such as Aletheia, StegExpose and StegSpy, that can detect hidden details without requiring manual investigation. A similar strategy of spotting the minute anomalies can also be applied to steganalysis of audio and text files, but in this chapter, we will limit ourselves to an image.

In 2002, steganalysis of images led to an international crackdown on the 'Shadowz Brotherhood', a child exploitation network spanning twelve countries and websites available in seven languages. Offenders had hidden child abuse material within seemingly ordinary image files, making them appear harmless to anyone but trained forensic eyes. Investigators used digital steganalysis techniques to detect irregularities in pixel patterns and file structures, particularly through LSB analysis, which revealed concealed data embedded in thousands of suspicious images.

Named Operation Twins, this investigation not only dismantled a major criminal organisation, but also demonstrated the need for powerful digital forensics, especially with the upcoming age of AI.

FORENSIC TOXICOLOGY 4

Poison in fiction

Poison is a covert weapon that gets the job done without the need for brute force or having to worry about cleaning up blood spatter from clothes. It indicates intent and deliberate planning. In real life, it makes the job of getting the killer behind bars a tedious affair, but in fiction, it makes for a great dénouement, with all the characters gathered around waiting for the final reveal.

Unlike a stabbing or shooting, where the murder weapon is out in the open, poison adds to the layer of mystery as the detectives slowly unravel the who and why of the killer. Many beloved murder mysteries are set in an era when forensics weren't an essential part of the criminal justice system, when murder by slipping poison into a drink or meal could easily be passed off as a natural death. Hence, it became a favoured messenger of death both in fact and fiction.

In season four of the TV show *Game of Thrones*, based on George R.R. Martin's *A Song of Ice and Fire* series, one of the most loathed characters, King Joffrey, is killed during his own wedding celebrations. He isn't stabbed with a sword, as one would assume in this fantasy world, but by drinking a glass of wine poisoned with 'the Strangler', described thusly by Prince Oberyn Martell, the Red Viper of Dorne, in the show: 'Dissolved in wine, it makes the muscles of a man's throat clench tighter than any fist. The victim's face turns purple as the little crystal seed from which his death was grown, and so they call it the Strangler.'

After drinking wine and eating pie, he suddenly begins to cough violently, and then his face turns red. He claws at his throat, gasping for breath, but nothing goes in, and he turns dark purple due to lack of oxygen in the blood. Blood leaves his nose and his life leaves his body.

The symptoms Joffrey suffers bear a striking resemblance to the effects of strychnine, a deadly alkaloid derived from the seeds and bark of the *Strychnos* genus of plants. In Martin's description, the poison is said to be extracted with a wash of lime before being treated with a rinse of sugar. Researcher Raychelle Burks, from Doane College in Nebraska, argues that lime might not be referring to the citrus fruit, but a compound derived from limestone, such as calcium oxide. Since calcium oxide plays a key role in the chemical process of extracting alkaloids like strychnine from plants, the fictional recipe is eerily rooted in real-world toxicology.

The works of Agatha Christie, the Queen of Crime Writing, include nearly 30 mentions of poisons, real ones, as plot driver and killer. Her hands-on experience with poisons and drugs as a pharmacist during the First World War gave her a deep understanding of poison doses and symptoms, which is quite evident in her stories featuring murders using poisons such as arsenic, thallium, cyanide, strychnine and many others – some of which we will examine in detail later in this chapter.

Fiction emulates life, and sometimes it's the other way round. It is 1600s Italy, and a woman walks to her local apothecary to buy some cosmetic healing water that can help cure the blemish that has been troubling her lately. She takes home a bottle of the healing water, and just as promised, the thing that was troubling her is gone. The blemish here was actually her husband, and the remedy came in the form of a mysterious illness that led to his death.

This was the story for many women in Italy when divorce wasn't an option, and the only way to freedom from an abusive husband was Giulia Tofana and the deadly concoction she prepared in her underground apothecary called Aqua Tofana. The colourless, odourless and tasteless liquid came in pretty bottles that purchasers could even display at their homes, where visitors would mistake it for holy water. Inside, however, was the unholy combination of arsenic, antimony and lead, possibly with the addition of mercuric chloride, according to some historical references. Aqua Tofana was a slow-acting poison that, depending on the dosage, allowed women to kill their husbands slowly over a span of several days or weeks. Death by this poison barely raised any alarms among family members and authorities, as the symptoms often mimicked those of common illnesses.

Giulia Tofana is believed to have been the daughter of Teofania di Adamo, a woman who was executed for producing and selling a poison later known as Acqua Tufània. Teofania may have been the original creator of this substance, and if that's true, Giulia's later involvement with the poison was her carrying on her mother's legacy and also her mission.

Giulia wasn't the only one in the poisoning game of the period. Mid-seventeenth-century Italy became home to a larger 'criminal magical underworld' with at least 200 people, including alchemists, wise women and astrologers, catering almost exclusively to an all-female clientele, mainly to murder cruel or unwanted husbands.

After Giulia's death around 1651, the baton was passed to Girolama Spara, a young accomplice of hers. Girolama, who was a widow of the renowned Florentine family, created connections within aristocratic circles, while the working-class arm of the poisoning ring was managed by Giovanna de Grandis. Working together, these women established a thriving business supplying Aqua Tofana to women seeking an escape from unhappy marriages across various layers of Roman society.

The fear of Aqua Tofana had spread so far and wide that Mozart, in his last days, believed he must have been poisoned with it.

However, everything came to an end in 1658 when the authorities caught the scent of these underground activities. How was the ring uncovered? Some say their secret was revealed to a priest during a confession, while others suggest a police agent, posing as a desperate husband, caught Giovanna de Grandis red-handed. Five ringleaders, including the two successors of Giulia Tofana and one of their clients, were executed in 1659.

Proof is in the poison

Alexander the Great had a dream. He wanted to turn the ancient land of Egypt into a hub of culture, religion and trade. His successor in Egypt, Ptolemy I Soter, soon set about transforming the city of Alexandria into a beacon of knowledge and intellect. The Library of Alexandria, possibly founded in 288 BC by Ptolemy I, stood as the greatest centre of learning in the ancient world. Within the walls of the library were manuscripts from every corner of the world, but this place was far more than a library. It was also a research centre and academy that brought together the finest minds under one roof. Among the many accomplishments of the city lies one of our areas of interest: the rise of toxicology.

Wherever there is power and politics, there is a fear of getting taken out by your enemy. In Alexandria, toxicology emerged alongside science and political life at royal courts, as knowledge of poisons was critical for both healing and protection against assassination.

In Ptolemaic Egypt, expertise in drugs and poisons made court physicians indispensable to the royal household. However, this same knowledge also aroused deep suspicion towards them, especially during times of political turmoil. For instance, Chrysippus of Cnidos, a physician, joined Queen Arsinoe I in a plot against King Ptolemy II Philadelphus. The King suspected that Chrysippus might try to poison him, even though the Hippocratic oath forbade doctors from doing such things. The King didn't trust him, so he had him executed solely based on a suspicion.

Despite the fear surrounding knowledge of poisons, the Alexandrian school became a centre for iology – a systematic study of poisons, antidotes and dosage – something that we know as toxicology today. Among the pioneers of this field was Apollodorus of Tarentum, also referred to as the 'father of iology', who drafted the treatise *Peri thêriōn* (On Poisonous Animals), where he systematically documented the current knowledge of poisons, venomous animals, treatments and antidotes.

Fast-forward a good few centuries to the Renaissance, when toxicology, like so many other fields, experienced a revival. Theophrastus Bombastus von Hohenheim, known as Paracelsus, now known as a Renaissance man, combined alchemy with medicine, trying different minerals and metals as remedies, something that his peers did not like.

As a medical man, Paracelsus experimented with metals like arsenic and lead, and he became especially known for promoting mercury as a treatment for syphilis. Yet earlier in his career, while working in the mines of Austria, he had witnessed the darker side of these same substances. He saw miners fall ill from the fumes of melted mercury and arsenic, their health corroded by invisible poisons. His experiences and experiments led to the establishment of the cornerstone of forensic toxicology.

In response to herbalists who criticised his treatment of syphilis, he wrote in *The Third Defence*: 'What is there that is not a poison? All things are poison, and nothing is without poison. Solely the dose determines whether a thing is not a poison.' It is a statement that has stood unwavering centuries later.

You might be thinking, surely not everything is a poison? Well, everything is indeed a poison; it's only a matter of dosage. Water? Yes. Oxygen? Yes. Coffee? Also yes.

Oxygen, a gas that is necessary to keep us alive, can damage lungs, cause seizures and even cause death if inhaled beyond a certain dose. Air is composed of approximately 21 per cent oxygen, and our bodies are tuned to function at these levels. The haemoglobin in our blood readily binds to the oxygen we breathe in and transports it to the rest of the body. Breathing in pure, 100 per cent oxygen overwhelms the body, as blood cannot carry away such high concentrations. The excess oxygen binds to proteins on the surface of the lungs, triggering oxygen poisoning. This can disrupt the central nervous system, blur vision, flood the lungs with fluid, and, in severe cases, lead to death.

Drinking water is essential for good health and skin and, most importantly, to keep our cells in shape and working. Dehydration is linked to issues ranging from temporary fatigue and dizziness to complications such as kidney damage and even death in severe cases. So, drinking lots of water is generally considered good advice. But how much water is too much water? Around 2.7 to 3.7 litres per day is considered a good amount of water for adults, but if one were to drink 6 litres at once, they might die from the kidneys overworking to get rid of the excess water and a drop in the body's sodium level, which is crucial for the proper functioning of cells.

Coffee contains caffeine, which in the right dose increases alertness and energy levels, providing you with enough of a boost to keep you going through a working day. A regular brewed non-specialty cup of coffee has around 113 to 247 mg of caffeine. For most adults, the FDA considers up to 400 milligrams of caffeine per day safe, which is roughly the amount in 2–4 350 ml cups of coffee, depending on the strength. But if one were to consume 1,200 mg or half a teaspoon of pure caffeine, they would experience toxic effects like heart palpitations and seizures, and if one were to ingest 10,000 mg or more, it could prove to be lethal.

Scientists apply the principle that the dose makes the poison to graph out dose-response relationships. The general trend observed is that the more you're exposed to a substance, the stronger its impact. So, experts map out this curve to understand where a drug heals and where it turns dangerous.

There are three different metrics:

Lethal dose (LD_{50}) is the amount of a substance that kills 50 per cent of a test population, usually expressed in milligrams per kilogram of body weight.

Effective dose (ED) is the amount of a substance that produces a specific effect, usually beneficial, like pain relief, but it can also refer to a harmful impact, depending on the endpoint measured.

Toxic dose (TD) is the amount of a substance that produces harmful effects. Comparing effective and toxic doses helps toxicologists and clinicians judge a drug's safety, since a dose that is effective at one level may become toxic at higher levels.

What is forensic toxicology?

Forensic scientists work with one central goal: to uncover answers that can shape criminal investigations and court cases. For a forensic toxicologist, every case starts with a single question: is poison at play? If the answer is yes, then come the expected questions: which poison? How was it delivered? What effects did it cause? And was the dose enough to kill?

A part of toxicologists' role also includes verifying intoxication and drug-related witness statements that come out during police investigations. If blood or urine samples are tested within twelve hours, they can confirm or deny claims such as someone blacking out on alcohol and not remembering what they did, or vouching that they haven't touched any drug for over a year, or that they lashed out under the influence of psychedelics.

Now, coming back to poisons, one of the first tests for poisons that gained celebrity status in the field of criminal investigations was the Marsh test.

The scene is a courtroom where John Bodle stands accused of poisoning his grandfather's coffee. The defence insists he is innocent, but the prosecution claims otherwise with science on their side, or so it seems. A British chemist, James Marsh of the Royal Arsenal at Woolwich, is called in to give expert testimony. Marsh applies Hahnemann's test, which involves adding a mixture of hydrogen sulphide and hydrochloric acid to the stomach contents collected from the victim. The sample produces a yellow precipitate, which is an indicator of arsenic being present. However, due to the instability of the sample, the yellow soon starts to fade, as does the jury's trust in his experiment. Without ironclad proof, Bodle walks free, only to later confess to the crime.

Marsh couldn't let this happen again: he spent day and night designing a test that was more stable. Eventually, he hit on a test for arsenic that wasn't plagued by the problems faced by previous precipitation tests. His test detects arsenic by treating a sample with sulphuric acid and zinc, which produces arsine gas if arsenic is present. This gas then passes through a heated glass tube where it decomposes into elemental arsenic, forming a black, shiny deposit called the 'arsenic mirror'. The appearance of this deposit confirms the presence of arsenic in the sample.

Once developed, it was time to use the test in real life. He didn't have to wait long; his test was first used in the trial of Marie-Fortunée Lefarge, accused of poisoning her husband with arsenic she had bought to kill rats. Initial attempts using the test failed, but when repeated by toxicologist Mathieu Orfila, arsenic was confirmed, leading to her conviction and public validation of the Marsh test. The irony here is that Orfila was originally hired by the defence.

Modern forensic toxicology

The first battle toxicologists need to conquer is proving if there is indeed poison present in the body before establishing it as a cause of death. To establish this, every toxicological examination involves four key stages. First comes detection, where screening tests reveal if any drugs or poisons exist in the samples. The second is identification, where specific physico-chemical tests are employed to conclusively identify drugs, metabolites or toxins. The third stage is quantification, aimed at accurately measuring the amounts of these substances. Before this comes the collection of the right specimens before the body undergoes an autopsy, so the pathologist must secure everything the toxicologist will need. What is collected might sometimes change depending on the case specifics, but some specimens are almost always gathered in every investigation:

Blood: The most important specimen for toxicology is blood. Ideally, two types are collected: central blood from the heart and peripheral blood from the femoral or iliac veins, found near the legs and pelvis, respectively. While heart blood is primary, it can be contaminated, making the smaller peripheral sample a cleaner source for analysis.

Liver: As the body's main site of drug metabolism, the liver often contains higher concentrations of drugs and their metabolites than blood, making detection easier.

Urine: While it is the most common sample for workplace drug testing or in patients who have been rushed to hospital with suspected poisoning, it's not always available post-mortem. Also, toxicology results don't reflect the drug at the moment of collection, since the body takes time to process and eliminate substances. However, if drug metabolites are found in urine, it is an indicator that they were in the body at some point.

Bile: When urine isn't available, bile from the gallbladder can serve as an alternative screening fluid. It often concentrates drugs like narcotics and benzodiazepines, making it important to collect all available bile.

Vitreous humour: The clear, gel-like fluid in the eye. The vitreous humour is a key specimen for post-mortem toxicology, especially for detecting alcohol, since alcohol levels in the gel are generally about 20 per cent higher than in blood. Its protected location also keeps it stable after death, making it a reliable specimen even when other samples have started to break down.

After the samples are collected, they go through a set of standard screening tests for the detection of poison. Immunoassays are the go-to screening tests for detecting drugs, using antibodies that only bind to specific substances in a sample. When the antibody binds to its target, the test produces a measurable signal, showing whether a drug is present or not. Results can be true-positive, false-positive, true-negative or false-negative. Any positive result needs a follow-up confirmation, usually done using spectrometry. This takes us to our next step of identification. Advances in analytical chemistry now allow toxicologists to combine this with the quantification step, the aim being to measure the precise amounts of these substances.

A tool commonly used for this process is gas chromatography combined with mass spectrometry, or GC–MS.

In GC, the sample is vaporised and carried through a long column filled with an inert gas. Different compounds move through the column at varying speeds, depending on their size, volatility and interactions with the column's coating. This step separates the drug or its metabolites from other substances in the sample.

This is followed by MS, where the testing process begins with the sample being converted into charged ions by bombarding it with electrons to create positive ions and fragments. These ions are then accelerated and passed through magnetic or electric fields, which deflect them based on their mass-to-charge ratios. The lighter or highly charged ions are deflected more than the heavier or less charged ones. A detector collects these ions and measures their abundance, producing a mass spectrum which acts as the 'fingerprint' of the molecules present.

For quantification, GC-MS uses the intensity of specific ion peaks corresponding to the compound of interest. By comparing the ion signal (peak area or height) to calibration standards of known concentrations, the amount of the compound in the original sample can be accurately determined.

The fourth and final stage is interpretation, where all the analytical findings from detection to quantification are evaluated in the context of the case. The toxicologist then compiles a clear, detailed report explaining the tests performed, what the results mean, and how they answer the questions posed by the investigating officer.

Prominent poisons and the chemistry behind them

Arsenic

The 33rd element, an unassuming-looking grey material much like its neighbouring metals and metalloids in the periodic table, has been responsible for the deaths of many a wealthy relative or abusive husband.

While the element is extremely toxic, its oxide form, arsenic trioxide (As_2O_3), also known as white arsenic, was popular among poisoners. Scientists who sampled the white, crumbly powder reported that it tasted metallic, pungent and sweetish. However, the victims of poisoning who survived to tell their tale did not recall any such odd taste, as white arsenic is easily masked in small quantities in food, hot beverages and alcohol.

Its sinister ability to blend in has been written about since the first century by Dioscorides, a Greek physician in the armies of the Roman Emperor Nero, who might have turned to arsenic for assassinating his stepbrother Britannicus.

Someone who definitely used arsenic to kill those who inconvenienced her was Nannie Doss, also known as 'The Giggling Granny', a serial killer who confessed to killing her children, husbands, mother and sister with a charming smile and often a chuckle. She poisoned her victims by lacing their food or drinks with arsenic, which was a common ingredient in household items like rat poison. She went on to kill at least eleven people over a span of two decades and never raised any suspicion due to the vague nature of arsenic poisoning symptoms.

She was caught after a doctor examined her fifth husband, Samuel Doss, for symptoms of foul play after his death. Autopsy and subsequent tests revealed fatal levels of arsenic in the body. Her motive? Many victims had life insurance policies, which she inherited after their deaths. She also killed for romance. When an investigator asked why she killed her husbands, she said with a smile: 'I was searching for the perfect mate, the real romance in life.'

Using arsenic to inherit wealth popped up in Philadelphia in 1938, home to the great arsenic poison ring. Stanley Phillips, a US Secret Service undercover agent who was investigating counterfeit money, untangled a murder-for-profit ring operated by two cousins, Herman and Paul Petrillo. The ring members, often posing as faith healers or witch doctors, targeted vulnerable and superstitious Italian immigrant women who sought spiritual and medical help to address marital or health issues. Once the bogus remedies failed, they suggested using a 'lafatura', or love potion, laced with arsenic that cost around $300, to 'cure' the spouse. Some wives were aware of the poison and OK with it, while others were tricked into believing that the victim died due to the love potion being quite potent, and their spouse failing to handle it. Either way, once dead, the victims' families or the gang itself, impersonating family, collected on life insurance policies.

In an era before forensics, arsenic barely got detected, as its symptoms are a great mimic of common health conditions and food poisoning. Most of the poisoning cases wouldn't even raise a red flag unless someone close to the victim became suspicious and requested an autopsy. Arsenic can enter the body through inhalation, skin contact and ingestion. When ingested, even a dose of 1 to 3 mg/kg of body weight of inorganic arsenic has proven to be fatal. On the other hand, the king of poisons was also used as a treatment for syphilis in small doses.

Once inside the body, arsenic, a neighbour of phosphorus, mimics the latter and replaces it in chemical interactions involving adenosine triphosphate (ATP), thereby disrupting vital energy-releasing processes that are responsible for fuelling our cells.

The earliest sign of a fatal arsenic dose is vomiting, which can happen within minutes or might even take hours, depending on the amount of food in the stomach. By the time the symptoms show up, much of the poison has already entered the bloodstream. The first bout of vomiting begins, and the victims start to feel parched despite drinking water. Next, the body tries to purge the toxin through the bowels. Diarrhoea typically starts around twelve hours after ingestion, turning watery and persisting until tenesmus sets in, where the bowel wants to empty despite having nothing in it. This causes extreme dehydration in the patient and convulsions, often leading to coma, followed by multiple organ failure.

The human body tries to expel arsenic as fast as it can, hence the extreme bowel movements and vomiting, but the liver and hair hold on to the metal for a long, long time. There is no hiding arsenic poisoning even after the victim's death, as the metal slows down the decomposition process, which results in unusually well-preserved corpses, perfect for a forensic scientist to run their tests. In cases of suspected arsenic poisoning, toxicologists conduct tests for total arsenic measurements, taking into account both inorganic and organic forms of the metal. Hair, urine and liver samples are used to carry them out.

While the Marsh test was where it all began with forensic tests for poison, it is no longer used as a test for arsenic in a medico-legal setting. Instead we use modern techniques such as molecular absorption spectrophotometry, which works on the principle that molecules absorb light at specific wavelengths, causing electronic transitions. The amount of absorbed light is directly proportional to the concentration of the molecules in solution.

Other analytical techniques include inductively coupled plasma mass spectrometry (ICP-MS), which uses hot plasma to turn elements in a sample into ions before a mass spectrometer measures those ions to find out which elements are present and how much of each there is.

Treatment of acute arsenic poisoning starts with stabilising the patient by giving fluids and correcting electrolytes. In severe cases, aggressive IV fluid therapy can save lives, while dialysis may be needed if the kidneys fail. The most important step is giving chelating agents that bind arsenic and help the body remove it as quickly as possible.

Dimercaprol (2,3-dimercaptopropanol, also known as British anti-Lewisite or BAL), 2-3-dimercapto-1-pr opanesulfonate (DMPS), or meso-2,3-dimer-captosuccinic acid (DMSA) are commonly used chelating agents for treating arsenic poisoning.

Strychnine

If the evil queen laced the apple with strychnine, Snow White would be on the ground convulsing within 30 seconds of taking a bite and would soon die from respiratory failure. Thankfully, she didn't.

As we have seen, strychnine is a highly potent bitter toxin, which is a naturally occurring alkaloid found in the seeds and bark of plants from *Strychnos* genus, particularly the *Strychnos nux-vomica* tree, whose disc-like seeds are sometimes called 'vomit buttons'. The poison was first isolated from Saint Ignatius' bean by chemists Joseph Bienaimé Caventou and Pierre-Joseph Pelletier in 1818. Sold in markets for centuries as a rodenticide, this poison has been the cause of death for many, real and fictional.

The Scottish-Canadian doctor Thomas Neill Cream, who later earned the moniker 'The Lambeth Poisoner', was a physician trained in midwifery who took it upon himself to murder prostitutes with strychnine and chloroform. Women who desperately needed an abortion came to him for medical assistance, but instead, he preyed on the vulnerable, whose precarious lifestyle provided them with little protection and possibly no one who would look for them if they were found dead. It is also suspected that he killed his wife, Flora Brooks, whom he was forced to marry after getting her pregnant. Though he abandoned her to move to London, Dr Cream later mailed her pills from across the Atlantic, claiming they would ease her ailments. He went on a murderous spree, killing sex workers across Chicago and parts of Canada, with strychnine pills, but it went almost unnoticed, as there weren't solid leads connecting him to their deaths.

Daniel Stott of Boone County, Illinois, also fell victim to Dr Cream's strychnine poisoning after he consumed the epilepsy medication prescribed to him by the doctor. At first, his death was written off as natural, but suspicions grew when Cream tried to blackmail the local pharmacist. The truth soon came out that Stott's wife, Julia Abbey Stott, who was also Cream's lover, had conspired with him to kill her husband. He was arrested but released after a decade, and he picked up where he left off, murdering four prostitutes in Lambeth in a span of seven months.

Dr Cream felt he was invincible, so he tried to extort London's high society, targeting a socialite, an MP and even a royal physician with anonymous letters accusing them of murder. Most brushed him off, but the letters landed on Scotland Yard's desk. He also taunted the police with letters containing chillingly specific details of the crimes and scenes. In the end, a handwriting match and his own arrogance sealed his fate. He was arrested, tried and hanged.

In Agatha Christie's first detective novel, the 1921 *Mysterious Affair at Styles*, Emily Inglethorp, the matriarch of a wealthy family, had recently married a man much younger than her. The family distrusted Emily's new husband, suspecting hidden motives behind the marriage, and her children, in particular, resented the potential complications it posed for their inheritance. One fateful night, sometime after taking her sleeping draught, Emily began seizing violently. The spasms came like a wave. One moment she would seize, and the next she would go limp. Then came one final convulsion where her back arched off the bed, and it appeared that the head and heels supported her body. After this, she was gone. Over the course of the next couple of hundred pages of the novel, Hercule Poirot, the hat-wearing, moustachioed Belgian detective, uncovered that the death was caused by a lethal dose of strychnine.

The description of Emily's death wasn't solely added for theatrical purposes. Agatha Christie, thanks to her experience working at a pharmacy during the First World War, gave an accurate description of the actual symptoms. Strychnine acts as a neurotoxin. As soon as it enters the body, it disrupts the activity of glycine, a neurotransmitter that tells the body when to switch the electric signals to muscles. With nothing to stop the electrical signals, the muscles undergo continuous contractions, resulting in spasms that prevent breathing.

The Mandelin reagent, a simple and effective spot-test for strychnine, was first introduced in 1883. This reagent is prepared by mixing concentrated sulfuric acid with ammonium vanadate, creating a powerful solution that interacts with alkaloid compounds to produce distinctive colour changes. Once this solution comes into contact with strychnine, it turns a purple colour, which changes to a striking rose-pink upon dilution with water.

Other analytical methods, such as gas chromatography–mass spectrometry (GC–MS), detect strychnine by first separating it from other compounds and then identifying it based on its unique chemical fingerprint.

As each compound leaves the column, it is bombarded with electrons in the mass spectrometry, which breaks it into charged fragments. These fragments form a distinct molecular fingerprint based on their mass-to-charge ratios.

Unfortunately, there is no antidote available for strychnine. The treatment is limited to good supportive care to help manage the symptoms. Intravenous fluids and benzodiazepines, which help block excessive activity of nerves, are given to help with the spasms. If the patient is unable to breathe due to muscle contractions, they are often intubated to facilitate breathing.

Cyanide

It is 9 September 2011. Roy Thomas, a 40-year-old resident of a village in Kerala, India, is throwing up his last meal inside his locked bathroom. His wife, Jolly Thomas, dials the number of a nearby carpenter to help get her husband out of the bathroom, as he isn't opening the door. The carpenter and a few neighbours arrive at the house and break down the door. They find him foaming at the mouth as they lay him down in his living room. They rush him to the hospital in the city nearby, but he loses his life on the way.

A post-mortem quickly determined cyanide poisoning, raising suspicions, but Jolly insisted that her husband committed suicide, which didn't sound far-fetched given Roy's alcohol habits, alongside his financial and relationship troubles. As a result, nobody pressed charges. However, a string of other sudden deaths, including those of Jolly's mother-in-law, father-in-law, and other relatives, including a two-year-old child, made the pattern too hard to ignore. It provoked calls for an investigation into Jolly's husband as six people related to her died unexpectedly between 2002 and 2016.

Investigators exhumed the bodies and sent them for forensic testing. The Central Forensic Science Laboratory in Hyderabad found no traces of cyanide in the remains of four victims, but the police tracked down the person who helped procure her murder weapon. Cyanide is an essential in jewellery stores in Kerala, as it is used to clean and polish gold. The police arrested her friend, who got cyanide from a local goldsmith, claiming that it was needed to kill big rats in her house. Later on, Jolly confessed to killing her victims with dog poison and cyanide.

Cyanide refers to the CN^- ion, which is made up of one carbon and three nitrogen atoms bound together by triple bonds. This highly reactive ion combines with hydrogen, sodium or potassium to form three of the most notorious poisons. Hydrogen cyanide is a volatile, colourless liquid that was deployed as a chemical weapon during the First World War. Sodium cyanide and potassium cyanide, by contrast, are white crystalline powders that, when ingested, react with stomach acid to release hydrogen cyanide gas.

Emperor Nero makes an appearance again. It is said that he extracted cyanide from cherry laurel water and used it to poison his enemies. In the Franco-Prussian War of 1870, Napoleon III urged his soldiers to smear their bayonets with the toxin, turning each blade into a chemical weapon. Although its reputation as a poison stretches back centuries, cyanide was first chemically isolated in 1782 by Swedish chemist Carl Wilhelm Scheele.

Cyanide, one of the fastest-acting poisons, kills by attaching itself to haemoglobin present in blood, which carries the poison to wherever the blood touches. It then binds cytochrome c oxidase, an enzyme in the mitochondrial electron transport chain that facilitates electron transfer to oxygen, a crucial step in cellular respiration. This stops the cells from using oxygen, which is essential for producing the energy required to power bodily functions. Organs with the highest oxygen and energy demands, like the brain and heart, are most severely affected, leading to symptoms such as headache, dizziness, seizure and loss of consciousness.

Cyanide can enter the body through unexpected sources. For instance, a two-year-old girl was brought to the paediatric emergency department by emergency medical services after developing vomiting, breathing difficulties and drowsiness about 30 minutes after eating a handful of fresh apricot seeds.

Matthew Crème, from Lancashire, was enjoying his bowl of cherries when he cracked a seed by mistake and discovered a centre that tasted like an almond. Curiously, he tried the same with two more seeds, and a few minutes after that, he started to feel ill. He developed a headache and began to feel tired. Googling his symptoms drew him a result of potential poisoning. He immediately called the 111 medical number and was admitted to hospital. Crème recovered, but a 2024 report details the clinical case of a woman who was rushed to the emergency department of a hospital after suffering acute cyanide poisoning after intentionally ingesting bitter almonds.

All three of these cases of cyanide poisoning were from unexpected sources. Bitter almonds, apricot seeds and cherry seeds contain a compound called amygdalin, a cyanogenic glycoside which, when chewed and ingested, is broken down by the gut enzymes to release hydrogen cyanide. Consuming even a small amount of these seeds can exceed the safe dosage of cyanide; for instance, six to ten bitter almonds can cause severe poisoning, and eating close to fifty could be fatal for an adult.

In cases of cyanide poisoning, doctors sometimes turn to activated charcoal to soak up the toxin from the stomach. Just one gram of charcoal can bind approximately 35 milligrams of cyanide, thereby reducing the amount absorbed by the body. Extra care is taken for patients who are vomiting, as protecting the airway becomes as crucial as feeding activated charcoal.

In the US, the go-to antidote for cyanide is hydroxocobalamin, which goes by the name Cyanokit. The standard dose is five grams administered intravenously over fifteen minutes, with the option for an additional five-gram infusion if needed. Hydroxocobalamin stops cyanide poisoning by binding to cyanide ions to form a non-toxic compound called cyanocobalamin, also known as vitamin B12, which is then excreted by the kidneys. Cyanide prevents the cells from using oxygen by binding with an essential enzyme. The antidote helps restore cellular respiration that is disrupted by cyanide.

The poison that has taken away many lives has also had a huge impact on how medicines are packaged. In 1982, the Tylenol Murders shocked the United States when seven people in the Chicago area died after ingesting Extra-Strength Tylenol capsules (a branded paracetamol) laced with lethal doses of potassium cyanide. Victims ranged from a twelve-year-old girl to adults, all dying suddenly after taking the common over-the-counter pain reliever. People were afraid to buy the trusted medication.

In response, Johnson & Johnson, the maker of Tylenol, swiftly recalled over 31 million bottles from shelves to protect consumers. This crisis propelled drastic changes in drug packaging: tamper-evident packaging became mandatory for over-the-counter medications, including sealed bottles, foil seals and blister packs, to ensure product safety. The poisonings also led to federal anti-tampering laws aimed at preventing similar tragedies.

Thallium

In a world older than the FDA, a wonder cream called Koremlu hit the US market in the spring of 1930, claiming to remove unwanted hair on the upper lip painlessly. The cream did indeed remove hair as advertised, but not in the intended way. One 26-year-old woman lost her eyesight, her teeth, and her ability to stand or walk due to Koremlu. She wasn't alone. Horror stories from women started pouring in of clumps of hair falling off their head and other parts, becoming paralysed and losing their eyesight. The painless hair-removal cream shared the main ingredient with rat poison – thallium acetate.

Thallium was discovered by both William Crookes in the UK and Claude-Auguste Lamy of Lille, France, who were the first to isolate the new element. William Crookes was asked to identify an unknown impurity present in sulphuric acid being used for mining. His routine tests even included the classic flame test – the same one many of us remember from high school – where a platinum rod is dipped into the metal-containing sample and then held in the flame to reveal its signature colour. The moment he had rod to flame, he saw a brilliant green colour which was unlike the green from barium or copper. His next stop was the spectroscope, an optical instrument that split light into its component wavelengths, allowing scientists to identify elements by their characteristic spectral lines. The spectral lines revealed that he had discovered a new element, which he decided to call thallium, derived from the Greek *thallos*, meaning a green twig.

Lamy was examining slime from a sulphuric acid plant at Loos, and also observed the green spectral line at the same wavelength as Crooke. He went on to extract thallium sesquichloride from the slime sample and then, via electrolysis, isolated the pure metal. At the 1862 International Exhibition, Crookes showcased a small sample of thallium, only to be upstaged by Lamy, who unveiled a gleaming ingot of pure metal and dismissed Crookes's powder as mere thallium sulphide.

While credit for discovery of the metal is still up for debate, Lamy did, however, recognise the insidious nature of the metal. He mentioned feeling unusually unwell while working in close proximity to the metal during the isolation process. The symptoms disappeared after he had finished with it.

The bluish-white metal (which turns grey in air) has earned the name the 'poisoner's poison'. The colourless, tasteless and odourless nature of thallium salt makes it very easy to mix into the food or drink of an unsuspecting victim. Unlike cyanide and strychnine, this is a slow-acting poison. The symptoms sometimes appear days after the actual exposure, making it difficult for experts to pinpoint when exactly someone was poisoned.

'I was convinced they had the wrong guy,' said Dubberly, whose mother, Peggy Carr, died after unknowingly drinking Coca-Cola poisoned with thallium nitrate by George James Trepal, a skilled chemist and programmer who seemed like nothing more than a quiet neighbour in their Florida community.

Why did he poison his neighbours? Apparently, he was driven to his wits' end by the Carr family, with their loud music, barking dogs and chaotic household. His first attempt to drive them out was a death threat pinned to their door, which they ignored. Then he did something sinister: he sneaked into their kitchen and laced bottles of Coca-Cola with thallium nitrate, resealing them with a capping device to avoid raising alarm.

Peggy Carr drank most of the poisoned soda, and within days, she started experiencing the symptoms of thallium poisoning. Her fingers tingled, the soles of her feet burned and her health rapidly declined. Even her hair fell out before she slipped into a coma and died weeks later. The husband and the sons fell ill but recovered after treatment and physical therapy.

Detectives became suspicious of George Trepal when, during a routine interview, he suggested that someone must want the Carrs gone. He was apprehended after investigators discovered a bottle containing thallium nitrate in his garage. Along with evidence that he had threatened the family before and the presence of a bottle-capping machine at his place, it pointed the finger of guilt his way.

Forensic toxicologists can detect thallium from samples of blood, hair and urine. In case of suspected poisoning where the victim is alive, the gold standard for thallium detection is a 24-hour urine test analysed by inductively coupled plasma mass spectrometry (ICP-MS) or atomic absorption spectroscopy (AAS). Despite the big fancy names, these instruments are nothing but more sensitive and sophisticated versions of the spectroscope Crooke used. With these methods, experts can measure even minute traces of thallium with precision, leaving little room for confusion with other metals. These techniques are also applied to post-mortem samples of tissues, including blood, brain, liver, kidney and hair from the victim.

Another tell-tale sign of thallium poisoning is the bayonet-like shape of hair shafts. Normally, during the growth phase, the hair root appears healthy and rounded. With thallium poisoning, the roots become damaged and the hair shafts take on an unusual, spear-like shape. When looked at under a light microscope, the hair roots also appear darker than normal.

Thallium is a potent poison because its positively charged ion closely resembles the essential potassium ion, allowing it to sneak into cells almost unnoticed. Once inside, however, thallium disrupts the normal functions of cells, particularly in the brain, nerves and muscles, where potassium is critical. By displacing potassium, thallium interferes with cellular processes, ultimately damaging the central nervous system and preventing hair follicles from producing new hair, resulting in hair loss throughout the body.

For thallium ingestion, repeated doses of activated charcoal (0.5–1 g/kg every two to three hours, up to four doses) can bind thallium in the gut and reduce absorption. The FDA-approved antidote, Prussian blue, which is marketed as Radiogardase, works even more effectively by trapping thallium in the intestines.

SOLVING CRIMES OF THE PAST AND PRESENT 5

Dr Temperance 'Bones' Brennan is a no-nonsense forensic anthropologist who sees a face in every skull. By studying bones, she can tell how someone walked and where they were hurt. Her very specific set of skills has led her to partner with FBI Agent Seeley Booth, and together, they solve crimes ranging from a chef brutally murdered in a meat grinder to a tragic love story where a woman learned that her vanished lover was not faithless but dead, and the gruesome work of a ritualistic, cannibalistic serial killer.

She isn't alone. Working alongside her at the Medico-Legal Lab at the Jefferson Institute is the Avengers of Nerds: a forensic entomologist who also specialises in soil, minerals and particulates; a forensic artist with expertise in facial reconstruction and digital imaging; a forensic pathologist and head of the Jeffersonian's forensic division; and a graduate intern, later a forensic anthropologist.

When it comes to creating a gripping narrative and awe for forensics, the TV show *Bones* gets a 10 out of 10. The science portrayed wasn't far from reality, but often fell prey to dramatisation, not unexpected from a procedural TV show.

Bones bear the weight of the human body for a lifetime. Hidden beneath layers of skin, muscle and sinew, they gather the marks of life's events, carrying a story written in a matrix of mineral and collagen. Forensic anthropologists, with their deep understanding of skeletal biology, human anatomy and physiology, can read the details left behind in the body's scaffolding to answer legal questions, help identify victims and uncover how they died. *Bones* does a good job of giving people a glimpse of what can be accomplished with forensic anthropology, with a few spices sprinkled on top for extra flavour.

For example, Brennan leans over a skeleton laid on the steel forensic table, glances at the bones for a few seconds and goes on to explain the gender, ethnicity and possible age of the person before they turned to bones. Forensic anthropologists can indeed glean these details, but it isn't by squinting and staring at the skeleton with a serious look. It takes time, patience and measurements.

Before we answer the questions of what bones can tell us, let's get a brief overview of what bones are.

Bones are made up of viscoelastic material, which is a perfect balance of springy and strong. Almost 70 per cent of bone by mass comes from inorganic materials, mainly calcium and phosphorus-based mineral hydroxyapatite and water, and the remaining 30 per cent is organic material, primarily collagen. While the minerals make the bone hard and load-bearing, a 100 per cent mineral bone would be extremely brittle. The collagen makes sure that the bone is flexible and able to absorb shocks without shattering into a thousand pieces like glass.

Bone tissues come in three variants: a harder type found on the outside called compact tissue, a sponge-like type found on the inside called cancellous tissue, and a smooth type found at the ends of bones called subchondral tissue, which is covered with another type of tissue called cartilage, a specialised, rubbery connective tissue. Apart from internal features that help bones resist impact, the thin yet tough outer membrane, the periosteum, wraps each bone to provide an added layer of protection.

Bone isn't just a rigid framework; it is a dynamic tissue that constantly responds to where we are in life. It reshapes itself with growth, shifting height and weight, and it bears the subtle marks of age, disease and injuries. Bones also hold within them the records of personal history and biological heritage, enabling forensic anthropologists to determine sex and ancestry.

If the skeleton presented to the forensic anthropologists is from an unknown source, meaning they aren't sure if the skeleton is human or not, their first task at hand is determining the species. Humans in their complete form with flesh and skin look very distinct from other species, but on a skeletal level, we do share similarities with other mammals. To identify what the skeleton is, experts focus on what is uniquely human: a very large case to hold our brain, a short and non-projecting jaw, and skeletal adaptations for walking upright.

The skull reveals some of the clearest contrasts. The cranial vault, a bone chamber that holds the brain, is very large compared to our body size and has a domed shape with a thinner outer layer of bone. Human skulls do not have the heavy sagittal and nuchal crests, the ridges of bone that give non-human primates their powerful bite and help four-legged animals support and lift their heads. Instead, humans have a large mastoid process that anchors the neck muscles needed to swivel and tilt the head in an upright posture.

Teeth provide another strong line of evidence. Human premolars and molars have a low and rounded design for an omnivorous diet, setting them apart from herbivorous teeth, which have high, ridged teeth, and carnivores, whose teeth are sharp and cutting. The case might get tricky if the only tooth available is a molar, because a human one looks very similar to a pig's fourth premolar. And how often does that happen in a case? One can assume not very often.

The spine also carries clear human signatures. Our vertebral column has an S-shaped curve to balance upright, and the vertebral bodies gradually increase in size from top to bottom to support more weight. Furthermore, anthropologists can distinguish between the skeletons of young people and adults based on the presence of unfused or partially fused growth plates, known as epiphyses, in the young ones of our species.

When determining the sex of a skeleton, the first choice is usually the pelvis, since it carries distinct markers of male/female dimorphism in humans. The female pelvis is generally broader, and this is most visible in the pubis. In women, the gap beneath the pubic bones is wider, since it surrounds the birth canal and has to make way for the baby's head to pass through during childbirth. Females also show a broader sciatic notch and have an extra feature called the ventral arc, which is usually absent in males.

Then we move on to determining the ancestry of the skeleton. Early methods of ancestry estimation were rooted in outdated race typologies, where anthropologists divided humans into rigid groups such as Negroid, Caucasoid and Mongoloid, often linking skull shapes or body measurements to stereotypes. Techniques developed by our old pal Alphonse Bertillon performed well with a limited population but showed inconsistent results when the scope was expanded. These methods not only reinforced racial stereotypes, but also attempted to neatly categorise humans, which is quite challenging because populations change over time due to the human tendency to migrate from place to place. Modern anthropologists avoid using 'race' as a scientific term due to its pseudoscientific origins.

To identify ancestry, experts once again focus on the skull. They begin by taking precise measurements of cranial dimensions such as length, width and facial proportions, and feed the data into computer programs like FORDISC or CRANID. These programs compare the data against global databases containing dimensions of thousands of skeletons in their collections. The results indicate how closely the unknown skull matches different groups and gives the result as probabilities, instead of relying only on an expert's eye.

The six-legged cleaning crew

In *Bones*, Jack Hodgins, the entomologist, often uses his army of beetles to strip away the decomposed tissue and prepare the bone sample for Dr Brennan to look at with a frown or for Angela, the forensic artist, to reconstruct the victim's face. While the show takes creative liberties with speed and the neatness with which the biological cleaning crew works, it draws on an actual scientific practice.

In the real world, museums and osteology labs routinely use dermestid beetles because they love to consume decaying flesh while leaving delicate bone intact, ideal for displaying skeletons for study and display. Major institutions like the Smithsonian, the Field Museum and the University of Tennessee maintain dermestid beetle colonies in dedicated spaces to clean and prepare skeletal specimens. A band of beetles, cared for by Hannah Trimble, a senior criminal justice major at the University of Tennessee at Martin, is even named after the Beatles: John, Paul, George and Ringo.

X-rays with scotch tape

It didn't take long after Röntgen's famous ghostly X-ray image of his wife's hand with a ring for forensic anthropologists to embrace X-rays as a window into bones, revealing hidden details without causing damage. X-ray images can reveal old injuries, diseases and subtle fractures, helping to distinguish whether trauma occurred before, during or after death.

If you have ever tried to peek into the X-ray machine in the hospital, you know what it looks like: a big white machine with a bed for a person to lie on and a projector-like device above it attached to a movable arm. Anthropologists use similar machines, sometimes of a smaller size, depending on the specimen being studied. But have you seen an X-ray machine that runs on Scotch tape?

In an episode of *Bones* in season six titled 'The Blackout in the Blizzard', the team find themselves stuck without power during a snowstorm while investigating a case. Waiting for the services to return wasn't an option, as a deadly pathogen was involved, and solving the case was the only way to prevent further infections.

How do you get an X-ray without electricity? Wendell, one of Dr Brennan's interns, recalls a clever trick Dr Brennan once wrote about. He rigs up a vacuum chamber into a makeshift X-ray by laying the bone sample on photosensitive plates and then stretching out six or seven parallel lines, where each line was made of two pieces of Scotch tape stuck together. He then pulls apart the tapes in a vacuum by cranking a handle. This releases enough energy to expose the plates and create an X-ray image of the bone. Unlike the show's more fanciful gadgets, like the laser 3D scanner or hologram machine, this setup is inspired by real physics.

Back in 2008, researchers from the University of California, Los Angeles, discovered that peeling Scotch tape in a vacuum at just three centimetres per second doesn't just produce visible light, but also generates enough X-rays to image a human finger. This effect is a form of triboluminescence, the phenomenon where certain materials release light when subjected to friction or mechanical stress from pulling, crushing or scratching. Scotch tape is one example, as are Wint O Green Life Savers (similar to Polo mints), which give off tiny flashes of light when crushed. The image from a Scotch-tape-powered X-ray setup might lack clarity, but when you are racing against time to stop the spread of a deadly pathogen, it is far better than nothing. These instances give us a glimpse into the fact that Dr Brennan and her adventures are loosely based on the life and novels of a real and respected forensic anthropologist, Kathy Reichs.

Also, let me put your mind at ease: no, you will not be exposed to X-rays when unwrapping a gift on your birthday or Christmas, as this effect is limited to a vacuum. Unless you are unwrapping gifts while spacewalking, it shouldn't be a problem.

Who died?

In early April 1997, a quiet rural area in upstate New York was shattered by gunshots. A local man known for his fiery temper attacked his daughter, who was strangely also his common-law wife, with a shotgun inside their house. Despite her injury, the woman managed to escape to her car. This time, the man didn't miss. He fired a fatal shot that ended her life. A neighbour, who was alerted by the victim's brother, called the police. Unfortunately, by the time the police arrived, the suspect had vanished into the dense woods behind his house, never to be seen again in the area, despite a massive manhunt.

Four years later, a hunter discovered a human skeleton hidden under leaves in the forest. Long after flesh decays, bones and teeth endure, carrying vital clues to a person's identity. Forensic osteology, a specialised branch of forensic anthropology, focuses on the study of human skeletal remains to aid legal investigations. By examining features such as age, sex and stature, and sometimes even evidence of trauma, forensic osteologists help reconstruct lives and uncover the circumstances of death.

Using the techniques we discussed in the previous section, the bone detectives confirmed that the skeleton was of a human male. Next comes age and stature estimation.

Fully fused bones confirmed that the human was an adult. Certain bones in our body change in a very predictable fashion with age, cranial sutures being one of them. The sutures are joints responsible for holding together different bones of the skull. In infancy, the cranial sutures are wide and open, allowing growth of the skull and brain. As we grow older, these start to get closer and closer, with complete fusion occurring above 40 years of age. The sutures on the skull discovered in the forest were mostly closed, indicating an age older than forty. To narrow it down further, they examined the end of the fourth rib, which changes shape in expected ways as people age. This rib suggested the person was likely between 43 and 55 years old.

Stature estimation is based on the well-established link between the sizes of certain bones and a person's height. In ideal cases, where a complete skeleton is found, forensic anthropologists will usually measure all the bones that contribute to a person's height and add them up, correcting for missing soft tissue and cartilage that decay after death, to get an estimate of the individual's actual living height.

Most cases aren't ideal: in such situations, experts have to make do with measuring whole limbs if lucky, or even bone fragments. When entire limb bones, like a thigh bone or shin bone, are available, their lengths are measured and then put into specific mathematical formulas that have been developed for different populations and sexes, based on analysis of large anthropological datasets. The formula calculates the estimated stature. Similar formulas also exist for bone fragments.

Using such calculations, the experts estimated that the living height of the male was 6 feet and 4 inches. The shotgun found beside the remains had its safety off and a spent shell casing, indicating a single shot. Skull fragments showed two exit wounds consistent with a shot fired under the chin, pointing to a self-inflicted death.

All this information, when put together, alongside the personal belongings and medical records obtained by the investigators, confirmed that the remains belonged to the man who fled to the forest after killing his daughter/common-law wife.

How did they die?

The method and mode of murder or death are crucial to catching killers. Trauma analysis examines changes or damage to bones caused by external forces, helping to answer questions about how and when injury or death occurred. However, certain caveats must be considered before treating a sign of trauma as evidence of violent or unnatural death. Not every fracture or dent is fatal. Over the course of life, humans collect marks of trauma on their bones from disease, injury or other causes. Identifying the true cause of death hidden among non-fatal traces requires careful, forensic attention to detail.

A great way of eliminating non-fatal traumas is to narrow down when the trauma occurred. Was it antemortem (before death), or was it perimortem (around the time of death)? In injuries that happened while the person was alive, the body has a chance to react to it and try to rectify the damage. One might see signs of healing, like new bone forming or remodelling, and rounded fracture edges. In perimortem cases, the breaks appear fresh and sharp, as no healing has occurred. So, it is implied that if a person lived long enough to heal after a trauma, the trauma was likely not the fatal injury.

The location of trauma on the body often dictates how fatal an injury it is, because different body regions have varying chances for survival. Some areas are more likely to be fatal. For example, skull fractures with radiating and concentric fracture patterns reflect forceful impacts that may have seriously damaged the brain, potentially causing death. Neck trauma from compression due to strangulation can cut off the airway or blood flow, leading to death. A fracture in the ribs can cause damage to the heart and kill someone.

On the other hand, trauma to the limbs, such as breaks or bends in the long bones of the arms and legs, is generally non-fatal. Sharp-force injuries like knife cuts or saw marks from dismemberment often occur after death, even if they happen ante- or perimortem, but away from vital areas, they are still unlikely to be fatal.

Sometimes trauma in non-fatal regions can become fatal depending on the mechanism of trauma. Trauma mechanisms are usually categorised into:

1. Blunt force trauma: This occurs when a force is applied slowly over a relatively large surface area, causing the bone to bend and permanently deform. Car accidents,

falls or being struck by an object like a club, hammer or fist are a few examples of such trauma. Strangulation also falls under blunt trauma, as it is a slow application of force through compression of the neck.

2. High-velocity projectile trauma: This involves a very rapid application of force over a relatively small surface area. The force is applied so quickly that the bone typically doesn't have time to bend or deform plastically. Bullets from firearms, shrapnel from a blast, or other very fast-moving small objects can cause such trauma. The impact of such force on bones is bevelling with an entry hole that is angled inwards and the exit hole is angled outwards, which can be an important clue when identifying the direction in which the projectile travelled. A high-velocity force also causes a greater degree of damage to the bones.
3. Sharp trauma: Caused by tools with a pointed tip or a beveled, angled edge, such as knives, saws, machetes or axes. Unlike blunt trauma, the force is concentrated on a very small area, making the impact highly focused despite often using similar motions. On bone, sharp trauma leaves distinct traces, including incisions, punctures or gouges, that can offer useful clues for determining the type of weapon used.

Now, coming to our premise of the mechanism of trauma deciding the likelihood of fatality from an injury, irrespective of the location. A blunt force trauma to the leg with a cricket bat will normally just break a bone, but in a high-force motor vehicle accident, the same impact can shatter the pelvis or cause internal bleeding, making it fatal.

A cut in the arm might not lead to death, but a bullet wound can cause severe trauma and can be fatal due to the extent of bone destruction, bleeding and damage to surrounding tissue.

Skilfully piecing together the signs of trauma helps forensic experts paint a picture of how a person must have died. But what happens to the bones after someone dies? Forensic taphonomy takes over at this stage. It is the science of understanding what happens to the body once life ends and the remains pass through nature's various temperature settings, encountering its different elements and living beings.

Originally, taphonomy was developed by palaeontologists to explain how fossils are formed, and now it is applied to modern forensic cases to understand how human remains change over time. As more time passes after death, clues start to be stripped away from the body, but time also writes new ones on the bones.

If a body is exposed to freezing temperatures, ice crystals can form within bone, damaging the cell structure and creating microfractures. A body buried in a shallow grave or left exposed shows signs of rapid bone erosion due to weather or animal activity. Even the burial soil's acidity can make or break bone preservation. In acidic soils, the mineral structure of bone is gradually dissolved, causing skeletal evidence to weaken and eventually disappear. By contrast, neutral or alkaline soils help stabilise bone and greatly improve the chances of long-term preservation. So, by examining the state of the bones, experts can deduce what the soil pH might have been, which is a crucial clue when trying to determine if the remains have been disturbed or moved.

To infer any kind of information based on the details collected, scientists often use the pool of knowledge from research conducted in both controlled and real environments. The former is taken care of in a lab, but the latter requires body farms.

As macabre as the name might sound, body farms offer forensic anthropologists the closest look possible at death in motion. These outdoor laboratories of death let scientists observe how nature takes its course on donated bodies, helping them better interpret taphonomic clues and ultimately give the dead a voice in the courtroom. Instead of guessing how a body might decay in a given environment, forensic experts can base their conclusions on decades of real-world data.

Need to study decomposition? Set up an experiment just as you would in a lab, documenting how heat, humidity, rainfall, insects, animals and soil chemistry transform a body over days, weeks and years. Want to simulate crime scenes? Body farms can help with that. By placing remains in shallow graves, under debris or inside vehicles, researchers can uncover how to locate hidden burials with ground-penetrating radar, cadaver dogs and soil analysis.

Have to create a taphonomic profile? These facilities can help recreate how trauma, weather and scavengers leave their marks on bone and tissue.

Strontium trail

The chemistry of the UK is different from that of North America. Every geographical region carries its own strontium isotope signature, shaped by local geological activities and water bodies. The chemical stamp doesn't stay limited to the earth, but travels up through the food chain, leaving its mark on every creature that calls the region home.

A person might migrate to different parts of the world throughout their life, but their bones act as chemical archives of where they spent their earliest years. Bones continually exchange strontium isotopes (^{87}Sr and ^{86}Sr) with the food and water of their current environment, so their composition reflects more recent residence. Tooth enamel, however, stops exchanging strontium after about age four, meaning its isotope ratio preserves the signature of early childhood. If the strontium ratios in a person's bones and teeth differ, it provides strong evidence that they moved to a new location after age four.

A great practical application of this was seen during the humanitarian initiative Operation Identification, where forensic anthropologists came together to solve the case of unidentified bodies found at the Texas–Mexico border. In the early 2010s, South Texas became a hub for migrant border crossings, and the increase in the number of people moving in also led to an increase in the number of deaths. With increased border crossings along established routes, migrants had to adjust their plans and travel through landscapes with unforgiving terrain and heat.

Forensic anthropologists from Baylor University and the University of Indianapolis began the painstaking work of exhuming unidentified border crossers buried in a local cemetery, and the remains were sent to the Forensic Anthropology Center at Texas State University.

Teeth and bone samples from the remains were analysed to identify where the individuals were from. One such case was the remains of a man, OpID 0381, recovered in Falfurrias, Texas. Isotopic analysis of teeth suggested that his overall dietary pattern was more similar to that of individuals raised in the US than migrant populations.

Strontium ratio data from his tooth enamel isotopes pointed to several possible childhood regions in the US, including California, Arizona, New Mexico, Minnesota and Wisconsin. Mexico and Central America were ruled out via oxygen isotope analysis (using the same logic as strontium), but the strontium ratio still fit with areas such as southern Guatemala.

The anthropologists proposed two possible explanations for OpID 0381's origins. One is that he may not have been a migrant at all but was mistakenly treated as one after death. Another is that he may have grown up in the United States, only to be deported and lose his life while trying to return.

In another case, police discovered a mutilated and decapitated corpse with body parts scattered along the banks of the Royal Canal in Dublin. Autopsy and primary anthropological analysis indicated that the victim was a male of African descent. The lack of a head and extensive damage made it really challenging for investigators to carry out visual identification. Fingerprint analysis also led nowhere. With traditional avenues closed one by one, investigators turned to strontium analysis.

Scientists used strontium isotope analysis on the femoral bone. By measuring strontium isotopes in different parts of the bone, they found that the older bone carried a signature consistent with hot, low-altitude coastal regions near the equator, such as parts of East Africa. The newer bone segment reflected the isotopic signature of drinking water in Ireland, indicating the victim had lived there for several years before death. This analysis suggested the victim originated from East Africa, likely Kenya, and emigrated to Ireland around six to seven years prior to his murder.

Now with a broader idea of the victim's identity, the police were able to calibrate their search. Soon, the investigators were testing the DNA of a child believed to be the victim's offspring. The DNA match confirmed the victim was 39-year-old Farah Swaleh Noor, a Kenyan, who had claimed asylum in Ireland.

This reveal led to the capture of the infamous 'Scissor Sisters'. The two siblings, Charlotte and Linda Mulhall, in March 2005 murdered and brutally dismembered the body of Farah, who was their mother's partner. Charlotte attacked Farah with a Stanley knife while Linda struck him with a hammer, and together they stabbed and dismembered him, placing the body parts in plastic bags and a sports bag before discarding them along Dublin's Royal Canal.

Charlotte received a life sentence for murder, Linda was convicted of manslaughter and given 15 years, and their mother was later jailed for aiding and abetting. The chemical imprint left behind by the 38th element of the periodic table led to the closure of one of the most gruesome cases in the country's criminal history.

Written in bones

A forensic anthropologist at the Smithsonian's National Museum of Natural History solved a cold case from the seventeenth-century colonial era, the death of fourteen-year-old Jane. The information hidden in fragments of bones was both shocking and enlightening. They found that Jane was butchered after her death, not as an act of hatred but as a last attempt at survival. The English settlers of Jamestown resorted to cannibalism to avoid starving during the winter of 1609–10, a time of extreme food shortage and drought. Only 20 per cent of colonists survived this winter, and the ones who did had to take extreme measures.

The remains were an unexpected discovery by Jamestown archaeologists, who had been working for two decades to uncover the lives of individual colonial settlers. The location of the find was a cellar room that also contained a butchered dog and horse.

Once the remains were in the custody of Smithsonian archaeologist Douglas Owsley, the investigation began. The start was rocky, as they had to make sense of an incomplete skeleton, with less than 10 per cent remaining. Examining the shape of the skull and the size of the tibia indicated that the remains were of a female, and her third molar and the growth of her shin bone revealed she was only about fourteen years old when she died. The skeletal evidence and the historical context painted a picture of a newly arrived English teenager who was just getting used to living in the New World, but alas, she didn't live to see the sun shining and the flowers blooming in the next spring.

So how do we know she was butchered for food? A seventeenth-century recipe involving brains – non-human ones, of course. They found multiple hesitant cut marks on the skull and jaw, and uneven chops that pointed towards an inexperienced butcher trying to remove the flesh from the face and brain for consumption. They also found scrape marks of a knife near the jawbone that might have been made to remove every last bit of flesh available.

Not far away, another dig led forensic anthropologists to the skeleton of a 400-year-old murder and cover-up.

Kari Bruwelheide, an American forensic anthropologist who also worked with Douglas Owsley, found a clandestine burial of a teenage boy in a cellar pit at an archaeological site in Anne Arundel County, Maryland. It wasn't a typical burial site – the remains were concealed beneath a layer of ceramic fragments, ash and animal bones, items not usually found with human burials, suggesting they were deliberately placed there to hide the body. They also found a milk pan placed on the rib cage, which had acted as the digging equipment.

The biological profile revealed the skull belonged to a Caucasian male of European descent, about fifteen to sixteen years old, whose spine and teeth bore signs of damage from hard labour or disease, quite common among bonded servants in colonial Chesapeake. They also found a coin in the grave dated 1664 and a window fragment, so the scientists placed his death between 1665 and 1675. Why such a long timeframe? Because it was this decade when new laws were enacted to prevent private burials, a measure intended to prevent employers from concealing abuse or deaths.

Unfortunately, the young boy secretly buried in the cellar did endure violence before his last breath. The boy's right wrist showed a fracture consistent with a defensive wound, suggesting he had tried to block a blow shortly before death. Based on these indelible marks left on the calcium remains of a human body, the experts concluded the boy was likely killed violently and buried in secret to avoid punishment over his treatment.

All sorts of details, from a tiny chip of bone to an odd-shaped dip in a skeleton's hip, can help science tell the stories of people who weren't able to tell us themselves.

ART CRIME 6

Forensic analyst Nicholas Eastaugh, with his eyes fixed on the lens of the stereomicroscope, is meticulously examining every millimetre of the canvas, searching for the perfect spot from which to take a sample. There it is, a bright white spot!

The analyst extracts a speck of white paint, smaller than the width of a human hair, and feeds it into a Raman spectrometer. The instrument shines a monochromatic laser on the paint sample, which interacts with the chemical bonds of different materials within the paint and exchanges energy, resulting in a shift in electron energy levels. This shift, known as the Raman effect, detects a molecule's electron cloud polarisability during vibration (every atom and molecule in this universe is always slightly vibrating), which results in unique Raman fingerprints or peaks in a graph, allowing experts to identify chemical and structural information about the sample being tested.

Within a minute, the machine spits out a graph with spectral lines, where each dip and peak tells everything one needs to know about the chemical makeup of the paint. The analyst identifies a couple of peaks corresponding to the inorganic compound titanium dioxide. These peaks raise a red flag because titanium white wasn't even around when this painting was supposedly made, and artists of that era typically used zinc oxide for white. Now that the peaks are here, it can only mean one of two things: the painter was a time traveller who could visit the future and procure the paint, or the painting is a fake.

In 2010, these tiny marks on a graph brought down the €16 million empire that Wolfgang Beltracchi and his wife Helen had built over the span of 25 years.

One of Beltracchi's forgeries, labelled as *Red Painting with Horses* by the German expressionist Heinrich Campendonk, was sold for €2.4 million at auction in 2006, a record price for a Campendonk in the German art market. Due to the exorbitant price tag, the buyer wanted to conduct thorough due diligence and asked Sophia, the director of a gallery in Geneva, to have the painting analysed forensically for authenticity. The painting ultimately ended up in Nicholas Eastaugh's London laboratory, where he discovered traces of titanium white in the zinc-oxide-based white paint used in the piece, evidence that landed Beltracchi in prison for a considerable time.

The international art market generated a staggering $64 billion in 2019 alone, and this enormous figure is expected to continue growing as demand for art and antiques remains high. Unfortunately, not every piece has been touched by the hand of a master. Some experts estimate that four out of every ten artworks in the global market (including those in museums and private collections) are fakes, while others believe the number is even higher. Whether it's four or five, it's still far more than it should be.

But how do forgers manage to evade the eyes of all the art experts? Beyond copying the artwork down to the tiniest detail, master forgers also nail every item on the authenticity checklist, leaving behind a breadcrumb trail of exactly what experts are looking for to ensure that a piece is the real deal. Thus begins the cat-and-mouse chase of forgers and art detectives outsmarting each other.

Science of forgery and art of detection

The smell and feel of the crisp white pages of a new book are almost intoxicating; if only we could preserve them as they were right off the shelf. But alas! As time passes, lignin, the glue that holds cellulose together, the primary component of plant pulp used to make paper, oxidises and breaks down into smaller organic molecules, causing the paper to turn yellowish-brown.

Art forgers who work with paper trick the viewer into thinking that the watercolour they just forged last week in their garage is from the Second World War era by expertly staining the paper with steeped tea or brewed coffee.

Back in the late 1900s, Tom Keating, a painter and art restorer, single-handedly flooded the British market with over 2,000 fake paintings, a phenomenon *The New York Times* loved to call 'watercolourgate'. Not only did Keating create immaculate forgeries of over 100 artists, but he also paid extreme attention to detail to make the watercolour paintings look aged.

Old paper not only fades to a yellow hue with age, but also develops rust-coloured specks, known as foxing, caused by oxidised metal impurities or mould breaking down the paper's structure. To replicate this effect, Keating would slightly dampen the paper, place it carefully on the table, then, with a flick of his hand, throw some instant coffee powder into the air. As the coffee particles slowly settled and stuck to the damp paper, they created tiny brown spots. Once dried, those coffee marks looked identical to foxing.

The naked eye might fall for this trick, but it's hard to fool forensic experts and their trusty radiocarbon dating machine.

Everything that is alive or was once alive is slightly radioactive due to carbon-14 (^{14}C), a radioactive variant of our run-of-the-mill carbon-12 (^{12}C). Trillions of high-energy particles from outer space, aka cosmic rays, hit the Earth every day. These rays convert the nitrogen (^{14}N) present in air into carbon-14, which reacts with oxygen in our atmosphere to become radioactive CO_2, which then gets absorbed by plants during photosynthesis, eventually passing on to other life forms via the food chain.

In the grand balancing act of nature, the ratio of carbon-14 to carbon-12, about 1 in 1.3 trillion, stays surprisingly steady. Living organisms constantly renew their radioactive carbon stocks through biological processes, which come to a halt after death. With no new mechanisms to refill, the ratio gets disturbed as ^{14}C begins to deplete at a slow but constant rate. This change in the ratios helps people in labs decipher the age of an object made of natural material.

Since paper, canvas, wooden frames and plant oils used in paints were all part of a CO_2-breathing plant, they can be easily carbon-dated. Using some clever maths involving a formula that combines the decay constant and the ratio of carbon-14 in the recent levels versus that in the sample, the forensic experts can find the true age of the artefact.

If the machine spits out ^{14}C levels matching the 1970s, then that 'sixteenth-century Renaissance painting' you bought from the suave man in the tailored suit is definitely not what it claims to be. But if your painting shows ^{14}C levels dating back 100,000 years, it's still a fake – just for a different set of reasons. For one, humans weren't painting on canvas with oils that long ago. Secondly, the materials are likely modern and synthetic, which can mimic the appearance of extreme age by showing extremely low ^{14}C levels.

Forgers like Eric Hebborn quickly caught on to this authentication technique. In his tell-all autobiography, *The*

Art Forger's Handbook, he mentions taking trips to auction houses where he would rummage through unsorted lots of old prints and drawings. He also loved using the endpapers of old books he bought from antiquarian bookstores. As a bonus tip for forgers, he advises them to look out for holes in old paper created by worms or insects, as they could poke a hole in their plan and get them caught.

Authenticators know forgers often use old paper to beat carbon dating, so they keep an eye out for ink patterns around the wormholes. In genuine works, these holes appear over time, after the artwork is complete, leaving the edges bare. In fakes, however, the paint often seeps into the hole or lines its edges, a clear sign the art came *after* the hole. Eric describes a crude but effective tip to avoid this issue: chew paper into pulp, press it into the hole, cover with blotting paper, and gently tap until it blends in.

Ken Perenyi, an art forger who was also a part-time art restorer, knew all too well the new tricks forensic scientists and art experts had up their sleeves. So he used this knowledge to fabricate evidence to keep up the facade that his fakes were actually real art.

Oil paintings take ages to cure or dry, and the slow process involves the paint components reacting with the oxygen in the air to form 3D molecular networks. This process causes the surface exposed to the outside world to stiffen, leading to the formation of a network of cracks, also known as craquelure, over a long period of time. If a painting were a fake, it would be devoid of cracks, as fresh oil paint is gooey.

After finishing the painting, Perenyi would let the paint cure either under sunlight or in an oven, then apply a coat of varnish mixed with glue and gesso. Once dry, the surface would turn brittle, much like cured oil paint, which could then be cracked by gently rolling a rubber ball on the surface. The artificially cracked network was enhanced by applying a dark pigment to it to simulate the accumulation of dirt on the surface over time.

He also found out that authenticators had resorted to examining the artwork under UV light, looking for the faint glow of fluorescence given out by naturally derived paint supplies like egg tempera binder, linseed oils and rabbit glue on frames. Organic molecules present in these materials degrade over time into smaller molecules often containing a molecular structure that, upon interaction with UV light, glows owing to fluorescence, which becomes stronger as the painting ages.

The presence or absence of a glow provides experts with enough clues as to whether the painting has been manipulated or repaired, because newer coats of paint either have very faint or no glow. To beat this test and make his freshly painted pieces seem old, Perenyi carefully swabbed out old varnish from the paintings he restored and reapplied it on the fakes to make them glow under a UV lamp.

Unfortunately, Perenyi could get away with such things because he never directly handled the sales of his work and operated through intermediaries. By the time the trail led back to him, the statute of limitations had already expired. So, does that mean forgers can just walk away from their crimes? Not quite. As I said, it's a cat-and-mouse game where we sometimes let the mouse run free just to understand its tricks. But now, let's dive into some hardcore science that's making it a whole lot harder for forgers to slip through the cracks.

Getting up close

Sometimes, identifying a fake requires something as simple as taking a closer look at the painting. Microscopy allows forensic experts to discern the tiniest details in the paintings. If an expert looked at oil paintings made by Perenyi under a microscope, they would soon realise that they were fakes by observing the cracks, especially what's inside the cracks. Usually, when a painting is left in the open, the cracks accumulate dirt over time, giving them a darker colour, but Pereyni's painting would show homogeneous dark spaces instead of gritty dust particles.

Forensic experts have found human hair, paintbrush bristles and even fingerprints on the painting surfaces. Once, the US faced an issue with a number of Jackson Pollocks whose provenance was unclear. Provenance is essentially a piece of paper that traces the life of an object from the moment it is created, mapping its journey through time and space, from the walls of the living room it adorned to the exhibition that showcased its presence, and finally to the attic where it stayed hidden for ages.

Thankfully, one such piece had a hair stuck in it. Comparing the DNA of the hair found in the painting with that of hair found in the suspected forgery studio revealed that the painting was indeed fake.

X-rays aren't just for looking through our skin and flesh to check for broken bones – they can also uncover what's hiding beneath a painting. X-ray photography is a non-destructive forensic technique that can penetrate low-density materials, such as paints and canvas, to reach areas where visible light cannot.

In an X-ray image, dense layers of paint show up lighter, which can be a sign of overpainting or hidden changes. These scans can uncover rough sketches or entire underpaintings lurking beneath the surface. According to art experts, originals tend to have chaotic underpaintings as artists often change their minds mid-stroke, wrestling with composition. However, in forgeries, the underlayers are usually too confident and clean, as the forger already knows exactly how the painting should end up looking. And for those trying to pass off fakes on reused canvases? X-rays can be a total nightmare, exposing hidden artworks that were buried under the outer forged layer.

Raman spectroscopy might have made headlines for helping catch master forger Wolfgang Beltracchi, but it's not the only trick up a forensic expert's sleeve that can reveal the chemical makeup of paint, varnish, metal sculptures and even ancient glassware. Another approach is X-ray fluorescence, and it works like this: when high-energy X-rays hit the surface of an artwork or artefact, they knock electrons out of the atom's inner shells. Electrons from higher levels then drop down to fill the gaps, and in doing so, they release energy in the form of X-ray fluorescence. The energy released is unique to each element, much like a signature, allowing experts to pinpoint exactly which materials are present.

These techniques can get most forgers into serious trouble, but sometimes they can also get someone *out* of serious trouble.

The man who fooled the Nazis

Han van Meegeren was an expert at faking paintings of the seventeenth-century Dutch master Johannes Vermeer and had made millions selling them to passionate collectors during the Second World War. One of his paintings was bought by a member of the Nazi top brass, Hermann Göring.

After the war, law enforcement in the Netherlands quickly caught up with van Meegeren, and he was put on trial for treason and accused of selling a national treasure to the Nazis. Realising that admitting to art forgery would carry a much lighter sentence than treason, van Meegeren decided to confess. He told the authorities that the painting he had sold was actually a fake. But no one believed him. To the detectives and art experts, it sounded like a desperate attempt to escape the far more serious charge. So he asked for his paintings to be tested in a lab.

One of the techniques used by experts was gas chromatography–mass spectrometry (GC–MS). In this method, a tiny sample of paint is taken from the artwork and fed into a GC–MS machine, which vaporises the sample and separates its components. As we saw earlier, it works on a principle similar to a simple school experiment, like when you draw a line with black ink on filter paper and dip the edge in water. As the water rises through the paper via capillary action, it carries the ink with it, gradually separating it into different colours – blue, green and red – based on how each component interacts with the water and its molecular weight. GC–MS does something similar, just at a much more precise and scientific level, allowing experts to analyse the chemical makeup of the paints and detect inconsistencies that could reveal a forgery.

During this test, they discovered Bakelite in van Meegeren's paintings – a synthetic polymer he used to accelerate drying and simulate the craquelure effect. However, Bakelite was not invented until 1907, centuries after the Dutch master had died. And that's how forensic analysis not only caught a forger, but also saved him from a treason charge.

Wine with tasting notes of World War

A great piece of art can make people feel a lot of emotions, and for some people, a great glass of wine does the same. This love for art and wine has created a billion-dollar market, and wherever there is money, there is crime.

Back in 2005, a retired FBI agent, Jim Elroy, was hired to solve the case of four eighteenth-century wine bottles that were sold for half a million dollars. The bottles in question were from the US President Thomas Jefferson's personal collection. Elroy's client had a suspicion that the bottles were fake and wanted him to investigate.

After asking around and following some leads, Elroy pinpointed his suspect, a man named Hardy Rodenstock, who discovered these bottles and set them up for auction. Unfortunately, Rodenstock couldn't be pursued with any solid evidence. Elroy was scouring through *Scientific American*, hoping to find forensic methods for testing the wine without opening the bottle. He struck gold when he came across French physicist Philippe Hubert's wine authentication technique.

As a result, Elroy flew to France with the bottles to see Hubert, who, after conducting a few tests, declared that the wine was made after 1945. How could he be so sure about its age? Hubert found traces of radioactive caesium-137 in the bottle, an isotope that does not occur naturally but was only created after the first atomic bomb went off.

The first atomic bomb testing happened on 16 July 1945, at Los Alamos, where the US scientists used the radioactive element plutonium in its core. This bomb led to an explosion that reached temperatures 1.5 times hotter than the surface of the sun. The catastrophic explosion changed the chemistry of our atmosphere forever by sending across it millions of radioactive caesium-137 atoms, a remnant of the atomic age that still lingers in our air.

The caesium-137 scattered throughout the planet's atmosphere is absorbed by the grape plants used to make wine. Since Hubert found traces of caesium-137 in the four Jefferson bottles, he could say for sure that the wine was made after 1945. Armed with this evidence, Elroy was finally able to take Hardy Rodenstock to court.

A similar technique of looking for chemical signatures of the atomic bomb is also used by art experts to spot counterfeits. Binders or mediums used for paintings come from plant materials that also absorb caesium-137 from the atmosphere. So, if someone claims that they are selling you an original Leonardo da Vinci painting from the 1550s but the painting is screaming caesium-137, then I'm sorry, but your Renaissance painting was made after 1945.

IS FORENSIC SCIENCE FOOLPROOF? 7

DNA blunders

It was a particularly warm day in Heilbronn, Germany, on 25 April 2007. Two police officers, Michèle Kiesewetter and Martin Arnold, parked their patrol car under the shade of a tree in Theresienwiese park. They were enjoying their lunch when suddenly a commotion broke out between two people behind the car. Before the officers could draw their weapons, the miscreants shot them both in the head at point-blank range.

A passerby called an ambulance immediately after discovering Michèle slumped in the driver's seat and her partner severely injured and covered in blood. By the time they were taken to the hospital, Michèle had succumbed to the gunshot. Martin survived, but he slid into a coma and was rushed to surgery.

This incident led to one of the largest criminal investigations in Germany's history. The police sprang into action in search of any evidence that could lead to the identification of the assailants. Weeks passed, but the investigation seemed to go nowhere. The investigators were eagerly waiting for Martin to come out of the coma in the hope that he would be able to provide some useful information about the attackers. Unfortunately, Martin's injury left him with no recollection of what the attackers looked like or what exactly transpired on that fateful day.

Although this door was closed to the investigators, the forensics department finally made a breakthrough, almost three months later. As per procedure, the different regions of the crime scene, in this case, the car, were swabbed to collect possible DNA traces left behind by the attackers. The samples were tested and run through Germany's central crime database. And there was a match!

The microscopic traces of DNA found on the rear passenger seat and central console of the squad car belonged to none other than the German police's white whale – the Phantom of Heilbronn.

The Phantom was an elusive yet prolific criminal who was a suspect in 40 crimes across Europe. She first came on to the German police's radar in 1993, when her DNA was found on the rim of a teacup that belonged to a 62-year-old woman who was strangled to death with a strand of wire taken from a flower bouquet in her living room. The police chased the clue but had to give up since there were no witnesses, fingerprints or description of what the killer might look like.

For the next decade and a half, the Phantom's DNA kept popping up in several locations across Europe. In 2001, her DNA was found on a kitchen drawer in the southwestern corner of Germany at the house of a 61-year-old antiques dealer who was strangled to death. A few months later, her DNA was found on a heroin syringe in Belgium, and two weeks later, DNA was collected from an abandoned biscuit found at the scene of a burglary, which also had traces of her DNA.

The Phantom was a ruthless and versatile criminal. Apart from being a suspect in six homicides, this femme fatale of Eastern European origin was involved in car theft, a school break-in and a robbery across the border, in France. She somehow managed to operate for years without being spotted once. No one had ever seen her, not even security cameras. And on the rare occasions someone thought they had, they would describe the features of a man.

Before we dive further into the escapades of this prolific criminal, let's address the elephant in the room: if no one had ever seen the phantom, how did the police know it was a woman?

To identify which sex DNA belongs to, scientists look for special parts of the DNA carrying the sex-typing markers, such as amelogenin, which is a type of protein that is essential for the development of tooth enamel. The gene responsible for producing amelogenin can be detected from DNA traces left behind by biological samples like blood, saliva, hair, tissue, bone and sweat.

Amelogenin-producing genes are found on both the X and the Y sex chromosomes, but they have different sizes and can be separated via electrophoresis. This is a laboratory technique used for sorting molecules like DNA fragments based on their size with the help of an electric field. The scientists add the sample to be tested into a special electrophoresis gel and pass current through it. This causes the molecules to move through the gel at varying speeds and directions, depending on their size, which leads to the separation. These separations appear as bands on the gel.

In this case, since the *AMELX* and *AMELY* genes are of different sizes, they can be separated and shown as two bands if the DNA is from a male, as males have both an X and a Y chromosome. If the DNA is from a female who has two X chromosomes, only one peak appears in the results.

Another method is looking for Y-STRs, or Y-chromosome short tandem repeats (see Chapter 3). These are repeated DNA sequences found only on the Y chromosome. The best part about this test: female DNA can't interfere with Y-STR analysis because women don't have a Y chromosome. So, any Y-STR data found can only come from a male; if it doesn't, then the DNA definitely belongs to a female.

Coming back to the German police and its hunt for the Phantom. Neither the crimes she committed nor the associates she worked with followed any discernible pattern. Her accomplices included Iraqis, Serbs, Slovaks, Romanians and Albanians who, even after getting caught and convicted, denied her existence. By 2009, almost €2 million were spent in her pursuit, and Heilbronn police alone racked up 16,000 hours of overtime and even announced a reward of €300,000 for any information on the Phantom ... only to realise that the Phantom never existed. She was nothing more than the ghost of a forensic mistake.

In March of the same year, investigators were trying to identify the burnt corpse of an asylum seeker. They conducted a DNA analysis, and lo and behold, the DNA belonged to the Phantom. But this was impossible. For one, the body was first discovered in 2002, yet the Phantom's wrath had continued for long after. And then there was the undeniable fact that the remains belonged to a man, since the asylum housed only male patients.

Upon following these red flags, the investigators came up with the theory that the Phantom DNA obtained from the crime scenes must have already been present in the cotton swabs used for collecting the DNA samples. All the DNA could probably be traced back to some unknown worker at a cotton swab factory. In the case of the Phantom of Heilbronn, a forensic slip-up may have cost time, money and manpower, but in some cases, the stakes can get much higher. It could rob an innocent person of their freedom.

The suspect who was never at the crime scene

Adam Scott, a man who was wrongfully convicted of a rape, had to spend five months on remand due to a forensic mix-up. LGC Forensics, Britain's largest private forensic science testing centre, was analysing the DNA sample collected from a rape victim in Manchester. The DNA turned out to be Adam Scott's. The police arrested him despite his never setting foot in the city and his phone records suggesting that he was in Devon, over 200 miles away from the crime scene, shortly after the attack.

Following Scott's arrest, Philippa Jefferies, Scott's solicitor, filed for a public inquiry into the mix-up at LGC Forensic. The forensic science regulator looking into the case found that Adam was the victim of a laboratory contamination from an unrelated incident he was involved in.

Adam's DNA was collected and extracted after a 'spitting incident' in Exeter. Instead of throwing away the plastic tray used for extracting Adam's DNA, the LGC technician reused the tray when analysing genetic material obtained from the rape victim. This avoidable human error led to the wrongful captivity of an innocent person for months.

Unlike the above two cases, there are scenarios where the DNA of a person completely unrelated to other people might still find its way into the sample despite following all the necessary protocols.

DNA transfers

As humans, we leave behind DNA every time we touch something, kiss someone, ride a train. Some of us more than others (more on that later). The fact that every contact leaves a trace presents the forensic experts with a tricky phenomenon of DNA transfer. It makes DNA profiling of a crime even more complicated due to the unavoidable possibility of some random person's DNA showing up in the analysis, despite taking all the precautions to avoid contamination.

A Silicon Valley millionaire, Ravish Kumra, was strangled to death by invaders who ransacked his house and also blindfolded and beat up his wife. DNA was collected from underneath the victim's nails in an attempt to identify the criminals. The DNA profile matched with Lukis Anderson, a homeless alcoholic whose name was in the criminal database due to a prior residential burglary conviction. His modus operandi was a perfect fit for this situation.

The police interrogated Anderson, but couldn't gather any information as he had no recollection of the night the invasion happened, partly due to his brain injury and partly because he spent the night detoxing at a local hospital after paramedics found him passed out drunk on the street.

So how did the DNA of a man who was nowhere near the crime scene end up under the fingernails of the victim? Though Anderson never crossed paths with the victim, their fingers came into contact with the same object but at different times.

The paramedics who responded to the invasion also treated Anderson. They clipped the same oxygen monitoring kit to the finger of each man, which led to some of Anderson's cells being deposited under the victim's. This is what is known as secondary transfer, a pesky little problem that raises a very pertinent question: if your DNA is found at a crime scene, does that mean you are guilty?

DNA transfer is often categorised into primary (direct) transfer and secondary (indirect) transfer. The former occurs when DNA is transferred directly from one person to another or to an object. For instance, a person touching a doorknob, sneezing on a surface, a victim scratching their attacker's skin, or bleeding on their clothing.

The latter happens when DNA is transferred from one person to an object and then to another person or object, without any direct contact between the two individuals, which is what happened in the case of Anderson.

The amount of DNA transferred also depends on a phenomenon known as 'shedder status' among scientists. This term is used to classify individuals based on how much touch DNA they leave behind when they make contact with a surface. The 'good shedders' tend to leave behind more DNA than the 'poor shedders', and the reason for such a difference could include whether they just washed their hands, the time of day, the type of surface being touched, and several other factors scientists are yet to discover. Hence, there is a chance that a crime scene might contain more DNA from an innocent person than from someone who was involved in the incident.

Does that mean DNA evidence is flawed? DNA is a powerful tool for solving cases, but it's not definitive proof of guilt or innocence on its own. Its true value lies in how it is interpreted within the context of the crime. Where, how and why can make all the difference.

Accusing someone because their blood was found on a murder weapon or finding DNA in a rape kit test (uncontaminated) is different from doing so based on a single hair found on a couch at the crime scene.

Forensic pseudoscience

The flipside of the rise in popularity of science in courts is the emergence of poorly tested, dubious methods being presented as forensic evidence. Forensic techniques are meant to be objective, repeatable, well-tested and backed by empirical observations. For instance, the uniqueness of fingerprints is supported by data collected over the years, and DNA analysis, which reveals a person's gender, has been established by numerous studies carried out at different times and in various parts of the globe. The same cannot be said for junk science, which is becoming increasingly popular in the criminal justice system.

'Junk science' sounds like science, but there is no empirical basis for the 'expert opinion'; it is 'subjective speculation masquerading as science', wrote M. Chris Fabricant, author of the book *Junk Science and the American Criminal Justice System*.

Hair comparison analysis, also known as microscopic hair analysis, is a technique where an examiner looks at hair strands under a microscope and compares visual features, such as colour, thickness, scale pattern, shaft diameter, pigmentation, etc. to determine if they could have come from the same person or animal. This is an analysis technique that is heavily reliant on the training and expertise of hair analysts.

In 2015, the DOJ, FBI, the Innocence Project, and the National Association of Criminal Defense Lawyers reviewed transcripts of over 500 trials where FBI examiners provided microscopic hair comparison testimony. They found that a shocking 257 out of 268 trials where the examiners testified against the defendant contained errors, which makes 96 per cent of the testimonies incorrect. The most harrowing aspect of this mistake was the impact on lives. In 35 death penalty cases that relied on hair analysis, errors were later found in 33 of them. Unfortunately, nine of those defendants had already been executed, and five more died while still on death row.

What if we could judge a person's criminal history based on how they respond to a call? Voice stress analysis, or VSA, is another technique with limited scientific basis for its results. VSA software is designed to measure subtle changes in voice patterns that might be caused (emphasis on the might) by the stress of trying to hide a deceptive answer. Certain VSA experts and software manufacturers claim that VSA detects microtremors related to stress, rather than directly detecting lies.

According to a 2008 study conducted by the National Institute of Justice, VSA results were no better than flipping a coin when it came to accurately detecting lies. In the context of drug use, VSA correctly identified only 15 per cent of individuals who lied about drug use and incorrectly classified 8.5 per cent of truthful people as deceptive.

Next up, we have bitemark analysis. This is a branch of forensic odontology that links a bitemark left on human skin to a suspect's teeth. The idea behind it is that just like fingerprints, teeth are thought to have unique shapes, spacing and wear patterns. When a victim shows a suspicious mark, say on the arm, shoulder or face, analysts compare it to dental impressions or photographs of a suspect's teeth.

One of the earliest mentions of such a method to identify a culprit was during the Salem witch trials in 1692. However, it only gained widespread popularity as a form of forensic evidence after the arrest of the infamous American serial killer Ted Bundy, where one of the key pieces of evidence presented against him by the State of Florida was bitemark analysis. Bitemark experts were having a field day during the 1990s as a result of this conviction, but the scientific basis of this technique is as robust as the evidence that the tooth fairy is real.

Ray Krone's story shows just how dangerously wrong bitemark analysis can be. When Kim Ancona was murdered in 1991, the perpetrator left bitemarks on several parts of her body, and the investigators called a forensic odontologist to check if they matched those of their prime suspect, Ray Krone. The expert compared the patterned marks on the body with the sample provided by the police, which was a piece of Styrofoam that Ray had bitten into. After inspection, the expert insisted that only someone with Krone's crooked teeth could have made the marks, and experts testified with absolute certainty that his bite positively matched.

Ray was convicted and sent to death row. A retrial and DNA testing in 2002 exposed the flaws in the case, which was largely built on the bitemark analysis. DNA analysis of the saliva collected from Kim's clothing did not match Krone's but pointed to Kenneth Phillips, a known crook with no crooked teeth, who was already in jail for another violent crime. After spending ten years behind bars, tried and tested forensic techniques helped set Ray Krone free.

Amanda Knox vs CSI blunder

American student Amanda Knox was convicted of murdering Meredith Kercher, a British exchange student, by Italy's justice system in a case built on DNA evidence.

Amanda Knox was an American student who moved to Perugia, Italy, also on an exchange programme. She moved into a villa at 7 Via della Pergola and shared an apartment with Meredith.

On 1 November 2007, just like many other nights since they met a week before, Amanda was at her Italian boyfriend Raffaele Sollecito's place. She returned to her house as usual, but something felt off. The front door was ajar, there were signs of disturbance, and Meredith's bedroom was locked. Even after her name was called, Meredith didn't answer. Assuming she was just sleeping late, Amanda went to take a shower, and that's when she noticed drops of blood in the bathroom sink. She started to ring Meredith's phone to ask if everything was OK: the phone rang and rang, yet no one picked up the call.

Amanda panicked and went to get her boyfriend; then they both informed the other residents of the villa that something was wrong. After not getting any response from Meredith, they broke her bedroom door only to find her lying dead with her throat slit and her body partially covered with a duvet. The police were immediately called.

The investigation began with the CSIs handling the crime scene and the detectives interrogating their prime suspects, Amanda Knox and Raffaele Sollecito. The police found it odd that Knox hadn't immediately reported the case and waited so long to raise the alarm. A 54-hour-long interrogation took place where she was questioned in Italian without a proper translator. She was also asked not to request a lawyer, as it would make things difficult for her. Four days later, she signed a statement in which she said she was at home when Kercher was killed and that it was Patrick Lumumba, the owner of a bar where she worked, who killed Meredith. Lumumba, Knox and Sollecito were all arrested.

Amanda soon retracted this statement and claimed that she signed the document due to the psychological pressure created by the police for telling the truth about her non-involvement.

Lumumba was released because customers at his bar recalled him serving drinks all night.

The trial is where forensics entered the scene.*

The CSI team collected over 480 separate traces, including DNA and hairs, with results compiled into a final report. Let's explore a few items on the evidence list and the prosecution's claims during Amanda and Raffaele's trial based on them:

1. The investigators found a kitchen knife in Raffaele Sollecito's apartment. Tests revealed Amanda's DNA on the handle and Meredith's on the blade. The prosecution called this the 'double DNA knife' because both the victim's and the suspect's DNA were present on it. They argued that this knife was the murder weapon and suggested that Amanda held it, which was why her DNA was on the handle, and attacked Meredith, which was why her DNA ended up on the blade. This was further supported by the findings that bloodstains found on the victim's mattress cover were compatible with the kitchen knife found.
2. On 2 November, the investigators photographed a bra clasp on the floor of Meredith's room, but it was not taken as evidence until 45 days later. The bra clasp carried her DNA as expected, but it also had a DNA

* This case went through multiple trials and appeals over several years. The discussion here focuses exclusively on the aspects most relevant to forensic evidence.

profile consistent with Raffaele Sollecito, as per the prosecutor's claims. They presented this as the physical evidence placing Sollecito at the scene, arguing that the DNA transfer meant he had participated in the assault by removing or handling the bra during the attack.

3. In the small bathroom shared by Knox and Kercher, mixed DNA traces of both women were detected on the sink tap, a box of cotton buds, the drain and the bathmat. Prosecutors claimed these traces indicated that Knox had washed up after the crime, suggesting an attempt to clean away blood or biological evidence.
4. Investigators found a clearly visible bloody footprint on the bathroom mat, while six more emerged only after applying luminol – a chemical that glows blue when it reacts with traces of blood, even if the area has been cleaned. A visual analysis of the bathmat print was found to be compatible with Sollecito's footprint. Of the hidden prints, one in Knox's bedroom matched her footprint, and two in the corridor were consistent with either Knox or Sollecito.
5. Beyond linking the prints to the suspects, prosecutors pointed out that most were only revealed with luminol, suggesting someone had tried to wash them away – a sign, they argued, of a deliberate effort to cover up what had happened.

Rudy Guede, an Ivorian acquaintance of Kercher, was arrested weeks later after vaginal swabs of the victim tested positive for his DNA, and a bloody handprint was found on a pillow at the scene. He claimed innocence, saying he was on a date with Kercher and returned from the bathroom to find an unknown man attacking her. Since he had prior known records, the police asked for a quick trial. Having no explanation in defence of his bloody handprint on the victim's pillow, Guede was convicted of murder and sexual assault and sentenced to 30 years in prison, which was later reduced to 16 for agreeing to a quick trial.

Despite being recorded on Skype telling a friend Knox had nothing to do with the murder, Guede later told police he saw her silhouette pass by the window before he left.

Holes in the claims

1. The kitchen knife evidence was riddled with problems. Chemical tests revealed no blood on the blade or handle, despite the fact that such tests are more sensitive than DNA analysis and can detect blood at dilutions of 1:10,000. Amanda's DNA on the handle was easily explained by her using the knife for cooking. On the blade, the DNA attributed to Meredith was so faint it was barely measurable, which was far below the threshold many labs consider a reliable conclusion.

Here is what Carla Vecchiotti and Stefano Conti, forensic genetics experts from the University of Rome who were appointed as independent court experts, had to say about the knife evidence:

> The electrophoretic profiles exhibited reveal that the sample indicated by the letter B (blade of knife) was a Low Copy Number (LCN) sample, and, as such, all of the precautions indicated by the international scientific community should have been applied.
>
> Taking into account that none of the recommendations of the international scientific community relative to the treatment of Low Copy Number (LCN) samples were

> followed, we do not accept the conclusions regarding the certain attribution of the profile found on trace B (blade of knife) to the victim Meredith Susanna Cara Kercher, since the genetic profile, as obtained, appears unreliable insofar as it is not supported by scientifically validated analysis.

The presence of DNA may have been a cumulative result of cross-contamination, collection and preservation errors. It was revealed that the knife was placed in an envelope that had previously been used to store gloves, and the officer who later handled the knife at police headquarters had been in Kercher's apartment, including Knox's room, earlier the same day. Both these incidents significantly increase the possibility of the sample being cross-contaminated.

Once the knife was at the police station, an officer noticed the knife packaging was not properly sealed, and so they had to close the gap with tape. Another officer later removed the knife and placed it in a non-sterile box sealed with tape. Careful collection and storage hygiene are crucial for the accurate analysis of forensic samples, and lapses like these can severely hamper the credibility of the results.

2. Next, we come to the mixture of both Amanda's and Meredith's DNA found in the bathroom. As we learned in the previous chapter, every contact leaves a trace, and humans shed DNA in the form of dead skin cells, hair and bodily fluids on a daily basis. Both women shared a small bathroom in the premises, so it's quite possible that mixed DNA traces of both women would be found on the sink tap, cotton buds, the drain and the bathmat.

Prosecutors argued that these samples showed Knox washed up after the murder, framing the intermingled DNA as evidence of her attempt to clean away blood and conceal the crime. However, this could have just been a byproduct of house-sharing.

A similar lapse of protocol was also observed when handling the bra clasp. First off, the bra clasp was collected after a month and a half, and not from the same location where it was photographed earlier. A video recording of the evidence collection process showed that the officers failed to follow basic protocol, such as using sterile gloves or changing them every time they touched a new item, or ensuring the evidence was securely stored and wouldn't drop on an unsterile floor. Considering that DNA from all three individuals (the victim and the suspects) was expected to be present in the room due to their regular presence before the incident, such mishandling created the risk of transferring low levels of DNA from one object to another.

3. The prosecution claimed that the bare bloody footprint found on the bathmat belonged to Raffaele and could not belong to Rudy, as the print was too narrow to be his. In reality, however, Rudy's feet were not bigger or wider than Raffaele's. Furthermore, an expert witness at Raffaele's trial testified that the bathmat print did not match the defendant's foot.

As for the shoe and footprints that appeared after the luminol test, they came back negative for blood. Luminol is not a confirmation test for blood, as it also glows blue in the presence of saliva, rust, cleaning liquid and even turnip. The defence argued this proved the traces were not blood and could be explained by false positives from cleaning agents, especially since Knox and Kercher shared the flat.

The case against Amanda Knox and Raffaele Sollecito completely hinged on DNA evidence. No other physical evidence or witness statements linked them to the crime scene. The prosecution's entire case came crumbling down the moment the defence and experts began to question the reliability of the DNA evidence. This exposed the existing notion that DNA is an infallible piece of evidence that will swoop in to rescue a sinking criminal case and ensure justice is served. DNA evidence is a highly reliable piece of evidence that has sent many perpetrators to prison, but it is also prone to contamination, misinterpretation or misanalysis. DNA has become a prisoner of its own accuracy and reliability due to a lack of understanding by the people using it that evidence needs to be analysed with context and subtext.

DEBUNKING A FEW FORENSIC MYTHS FROM TV SHOWS

8

Contrary to what TV shows would have you think, a criminal investigation isn't usually a high-stakes, fast-paced procedure that catches murderers and serial killers every week. In most cases, it's a time-consuming process where every expert sticks to their lane and no one is a master of everything. Students enrol in criminology education hoping to become homicide detectives who can piece together a puzzle and solve a case, but most likely, detectives end up investigating simpler crimes, such as real-estate fraud or bike theft. Apart from the entire premise of criminalistics, here are my top four forensic science myths that need debunking.

You cannot expedite a forensic test based on its urgency

In TV shows like *NCIS* and *Law and Order*, a crime occurs, the CSI arrives at the location wearing a formal suit – not Personal Protective Equipment – and an ID hanging from their neck with absolutely no regard for cross-contamination.

They then pass some snarky comments to the police, collect the DNA, and the countdown begins. Every case is an emergency, and every case needs an expedited DNA analysis treatment. The investigators ask the forensic experts to carry out the DNA analysis as fast as they can, which can range from an all-nighter montage to the expert just inserting an ampoule of the sample into a machine and the DNA results coming out within minutes, right on time as she finishes explaining the process to one of her colleagues.

In real life, these DNA analysis results can range from days to even months based on the urgency of the case, backload, complexity of analysis, and non-technical reasons like administrative and logistical delays. CSI shows create an illusion that every police department has a dedicated forensics lab, which is not the case; only the larger departments with sufficient budgets can afford to set up a lab. So, many cases are sent to the limited number of police forensic labs and even external testing facilities. This demand often exceeds lab capacity, creating significant backlogs that delay processing.

DNA analysis is a complex multistep process; to speed it up, samples are often processed in batches, and every batch analysis can take upwards of a couple of weeks. Once the results are out the data is interpreted by an expert and reviewed by a second expert to ensure the highest level of accuracy. This entire process can take up a few months. This doesn't mean that cases aren't expedited, or rapid DNA analysis is a myth. Techniques like capillary electrophoresis can provide DNA results in just a few hours, but they aren't as commonplace as portrayed in TV shows.

Videos cannot be zoomed in and enhanced

The lead detective and his team of subordinates are staring at a screen with serious intent, trying to figure out the numberplate from a blurry image, then the chief instructs the person operating the system to 'zoom and enhance'. Bam! All of a sudden, we are much closer to the number plate, which in real life would make the already blurry video even more pixelated, but in this case, the picture becomes clear and reveals new details.

Recent advances in AI and image processing have enabled improvements in image quality by reducing noise and sharpening details, but this is not a magic solution. Firstly, image-editing software analyses the pixels around the selected area and intelligently reconstructs the missing portion. However, these techniques cannot create details that were never captured by the camera sensor. Scientists were able to recreate the process of detecting a person from looking at a reflection of a person in someone's eye, but the original picture was shot on a high-resolution 39-megapixel camera, which most CCTV and home security cameras are not. The spontaneous enhancements seen in TV shows where unreadable licence plates or blurry faces become perfectly sharp are still beyond current real-world capabilities.

Gun shell casings cannot be picked up with a random pen

In the TV show *The Flash*, Barry Allen, when he is not running around at superspeed and catching bad guys as the title character, helps the Central City Police Department catch criminals as a forensic chemist. In one episode, he reaches the crime scene, finds a bullet casing, dons his gloves, asks the police chief for his pen, uses it to pick up the gun casing and places it in a ziplock bag. If an actual CSI did this, it would derail the entire investigation. Why? The primary purpose of taking and putting it in a secure evidence bag is to prevent cross-contamination from external elements, which could compromise forensic analysis. Using a pen, a personal belonging that may well be contaminated with the owner's DNA and fingerprints, to collect evidence completely defeats the purpose of wearing gloves and packaging it in an evidence bag.

Investigators use sterile gloves and tools when handling evidence. Each evidence item is packaged separately to prevent cross-contamination, and bags are sealed with tamper-evident seals. Proper labelling is used to maintain integrity and ensure the chain of custody.

There's no all-knowing forensic genius behind every case

This book will hopefully have given you a glimpse into the vastness of the field of forensic science. There are hundreds of additional techniques and specialisations that we couldn't cover, as that would have made for a book five times as long. It is not humanly possible for someone to know the nitty-gritty of every forensic technique and carry out every step from evidence collection to analysis to interpretation.

Forensic investigations are a highly collaborative process, where everyone has a defined role and clear protocols to follow. There is a clear chain of custody, data sharing and review in place to minimise bias and misinterpretations. Many roles are specialist, not covered by a single individual.

Forensic science has proved incredibly useful for detection and the identification of criminals. But it has its limitations. The public and the police need to be aware of these to avoid misusing this powerful tool.

INDEX